Knowledge Management and Learning Organizations

Carolina Machado · J. Paulo Davim
Editors

Knowledge Management and Learning Organizations

Springer

- Keep at the forefront of innovative theories, models, processes [illegible] as well as the most recent research activities relating to knowledge management and learning organizations.
- Advance our understanding of [illegible] knowledge management [illegible] zational issues.

Organized in seven chapters, Knowledge Management and Learning Organizations looks to discuss in the first Chapter "Factors Affecting Knowledge Management and Learning: Exploring the Role of Diversity, Inclusion and Organizational Citizenship Behaviour" while the second focuses on "Knowledge Co-Creation". The third Chapter discusses the relevance of "Knowledge Management: To Share or not to Share?" and the fourth speaks about the "Knowledge Transfer: An Emerging Element of a Learning Organisation in Family Businesses." The fifth Chapter focuses on "Corporate Memory Systems: Key for Experienced Based Management", the following deals with "Knowledge Management: Looking for Success Profiles" and finally the last Chapter covers "The Misconception between Organizational Learning and Knowledge Management."

Nowadays, the interest in this subject is evident for many types of organizations, namely, important institutes and universities all over the world. Conscious of this importance, the present book looks to provide a support for [illegible] as well as those who are operating in the [illegible] and strategies related to knowledge management and learning organization issues.

The Editors acknowledge their gratitude to [illegible] their professional support. Finally, we would like to thank the chapter authors for their interest [illegible]

Braga, Portugal

Aveiro, Portugal

J. Paulo Davim

Editors and Contributors

About the Editors

Carolina Machado University of Minho, School of Economics and Management, Department of Management, Campus Gualtar, Braga, Portugal. She received her Ph.D. degree in Management Sciences (Organizational and Politics Management area/Human Resources Management) from the University of Minho in 1999, Master degree in Management (Strategic Human Resource Management) from Technical University of Lisbon in 1994, and Degree in Business Administration from University of Minho in 1989. Teaching in the Human Resources Management subjects since 1989 at University of Minho, she is since 2004 Associated Professor, with experience and research interest areas in the field of Human Resource Management, International Human Resource Management, Human Resource Management in SMEs, Training and Development, Emotional Intelligence, Management Change, and Knowledge Management and Management/HRM in the Digital Age/Business Analytics. She is Head of the Human Resources Management Work Group at the School of Economics and Management of University of Minho, Coordinator of Advanced Training Courses at the Interdisciplinary Centre of Social Sciences, Member of the Interdisciplinary Centre of Social Sciences (CICS.NOVA.UMinho), University of Minho, as well as Chief Editor of the International Journal of Applied Management Sciences and Engineering (IJAMSE), guest editor of journals, book editor and book series editor, as well as reviewer in different international prestigious journals. In addition, she has also published both as editor/co-editor and as author/co-author several books, book chapters, and articles in journals and conferences.

J. Paulo Davim University of Aveiro, Department of Mechanical Engineering, Campus Santiago, Aveiro, Portugal. He is a Full Professor at the University of Aveiro, Portugal. He is also distinguished as honorary professor in several universities/colleges/institutes in China, India and Spain. He received his Ph.D. degree in Mechanical Engineering in 1997, M.Sc. degree in Mechanical Engineering (materials and manufacturing processes) in 1991, Mechanical Engineering degree (5 years) in 1986, from the University of Porto (FEUP), the Aggregate title (Full Habilitation) from the University of Coimbra in 2005 and the D.Sc. (Higher Doctorate) from London Metropolitan University in 2013. He is Senior Chartered Engineer by the Portuguese Institution of Engineers with an MBA and Specialist titles in Engineering and Industrial Management as well as in Metrology. He is also Eur Ing by FEANI-Brussels and Fellow (FIET) of IET-London. He has more than 30 years of teaching and research experience in Manufacturing, Materials, Mechanical and Industrial Engineering, with special emphasis in Machining & Tribology. He has also interest in Management, Engineering Education and Higher Education for Sustainability. He has guided large numbers of postdoc, Ph.D. and master's students as well as has coordinated and participated in several financed research projects. He has received several scientific awards and honors. He has worked as evaluator of projects for ERC-European Research Council and other international research agencies as well as examiner of Ph.D. thesis for many universities in different countries. He is the Editor in Chief of several international journals, Guest Editor of journals, books Editor, book Series Editor and Scientific Advisory for many international journals and conferences. Presently, he is an Editorial Board member of 30 international journals and acts as reviewer for more than 100 prestigious Web of Science journals. In addition, he has also published as editor (and co-editor) more than 200 books and as author (and co-author) more than 15 books, 100 book chapters and 500 articles in journals and conferences (more than 280 articles in journals indexed in Web of Science core collection/h-index 59+/11000+ citations, SCOPUS/h-index 63+/14000+ citations, Google Scholar/h-index 81+/23000+ citations). He has listed in World's Top 2% Scientists by Stanford University study.

Contributors

Nick Chandler Budapest Business School, University of Applied Sciences, Budapest, Hungary

Diana Fernandes School of Economics and Management, University of Minho, Braga, Portugal

Delfina Gomes School of Economics and Management, University of Minho, Braga, Portugal

Felicity Kelliher School of Business, Waterford Institute of Technology, Waterford, Ireland

Octávio Lopes Lisbon School of Economics & Management, Universidade de Lisboa, Lisbon, Portugal

Carolina Feliciana Machado School of Economics and Management, University of Minho, Braga, Portugal;
Interdisciplinary Centre of Social Sciences (CICS.NOVA.UMinho), University of Minho, Braga, Portugal

Rita Milhazes School of Economics and Management, University of Minho, Braga, Portugal

Deirdre O'Donovan Munster Technological University, Cork, Ireland

Michael O'Meara School of Business, Waterford Institute of Technology, Waterford, Ireland

João Leite Ribeiro School of Economics and Management, University of Minho, Braga, Portugal;
Interdisciplinary Centre of Social Sciences (CICS.NOVA.UMinho), University of Minho, Braga, Portugal

Maria João Santos Lisbon School of Economics & Management, Universidade de Lisboa, Lisbon, Portugal

Yasemin Sen School of Business, Istanbul University, Istanbul, Turkey

Factors Affecting Knowledge Management and Learning: Exploring the Role of Diversity, Inclusion and Organizational Citizenship Behaviour

Deirdre O'Donovan

Abstract This chapter explores the impact of three distinct, but related, factors on Knowledge Management and Learning in organizations, specifically Diversity, Inclusion, and Organizational Citizenship Behaviour. The chapter begins by overviewing each of the factors, and other relevant concepts, before briefly discussing other organizational benefits associated with the three factors. Finally, the potential relationship between these factors and Knowledge Management and Learning is discussed.

1 Overview of Relevant Factors and Concepts

The three primary factors of interest for this chapter are:

1. Diversity
2. Inclusion
3. Organizational Citizenship Behaviour.

First, it is important to note that Diversity and Inclusion are terms that are often used interchangeably, but are not the same. Simply, 'Diversity' means differences (O'Donovan 2019), whereas 'Inclusion' is a method, or approach to diversity, by which differences are harnessed to the benefit of all.

Many definitions of diversity exist. Diversity can be defined concisely as all the ways in which individuals differ from each other (Joplin and Daus 1997), or as referring to:

> The similarities and differences in such characteristics as age, gender, ethnic heritage, physical abilities and disabilities, race and sexual orientation among the employees of organizations (Griffin and Moorhead 2006: 31).

Such definitions highlight the simplicity, and inevitability, of diversity. If diversity simply refers to differences between people, and people are generally accepted to be different in many ways, then diversity is ever present. This simplicity, however,

D. O'Donovan (✉)
Munster Technological University, Cork, Ireland
e-mail: deirdre.odonovan@cit.ie

C. Machado and J. P. Davim (eds.), *Knowledge Management and Learning Organizations*,
https://doi.org/10.1007/978-3-030-71079-8_1

masks a level of complexity associated with diversity. One element of complexity lies in the fact that diversity relates to both visible and invisible characteristics (Morley et al. 2004). Visible characteristics concern elements of an individual's make up that can be easily "seen" or identified, for example, skin colour, height, gender (physical presentation), accent or mannerisms. Invisible characteristics concern those elements of an individual's make up which cannot be "seen", or are more difficult to quickly discover, for example, gender identity, sexual orientation, belief system or family status.

As mentioned earlier, diversity is ever present. If diversity is ever present, it follows that organizations are, and always have been (to varying degrees), diverse. Even organizations that appear, or appeared, to have quite homogenous workforces, are inescapably diverse. If, for example, an organization had a predominantly white, male workforce, those men would be diverse, different, from each other. They would differ in terms of religious beliefs, sexual orientation, education, work experience, family status, culture, politics, etc. There has been, however, a history of assimilation cultures in some organizations, and of discriminatory hiring practices in others.

An assimilation culture is one in which employees are expected to behave in similar ways. This goes beyond following the rules of the organization, and relates more to adhering to norms and social cues. Essentially, employees leave 'themselves' at the door, meaning they mimic the behaviours of all other employees, minimizing expressions of differences. In such scenarios, diversity, what makes people unique, is shielded. As a result, all employees seem to be 'the same' (even though they are, of course, not). This type of environment may seem easy to monitor and control, however, the organization will typically find it difficult to be dynamic and flexible, and may have high levels of turnover, or low levels of engagement. They may also find it difficult to expand their attraction net to target employees from different talent pools, as such employees are unlikely to stay in an assimilation organization.

Regarding discriminatory practices resulting in seemingly homogenous work groups, these can be intentional or unintentional, and vary in format. Through history there have been incidences of blatant discrimination, some of which are well known, including, for example, the signs in England (Bagley and Abuaker 2017) or Australia (Stratton 2008) stating "No Irish Need Apply" or "No blacks, No Irish". In South Africa, the well documented Apartheid regime dictated what jobs were for "whites" and what jobs were for "blacks", or saw black employees experiencing inferior pay and conditions (Gradín 2018). In Ireland, the Marriage Bar prevented women from continuing in paid employment (Mosca and Wright 2018), particularly in education (Redmond and Harford 2010) and the public sector workforce, once they had married. Beyond such blatant practices, many other issues can result in discriminatory hiring, promotion and retention practices. While full discussion of this is beyond the scope of this chapter, there are a number of examples. Organizations may be using terminology in recruitment advertising that is ambiguous or causing potential candidates to self-select out. An example is a job advertisement that says the job is suited to 'flexible' candidates. What does it mean to be 'flexible'? Does it mean that the hours worked will vary each week, the times worked will vary, or that you may be called in on short notice? The organization may simply mean that you need to be able to adapt to different types of customers or customer needs, or to be able to think on your feet to problem solve, but, candidates with those abilities may interpret flexibility differently and so not put themselves forward for the position. This may mean missing out on the attraction of employees who have, for example, childcare or elder care commitments, educational commitments, or who do not have easy access to transport. Furthermore, if interviews and other selection tests are not firmly focused on assessing which candidate best matches the requirements set out in the Job Description and Person Specification, the potential for bias in the selection process exists. People often possess biases, which can be subconscious, meaning the person may not be aware of their own biases. Consequently, hiring managers may hire people similar to them ('Similar to Me' or 'Affinity' bias), rather than a candidate who is readily identifiable as different to them in some way.

In recognition of these and other issues, organizations have become more cognizant of diversity. Some organizations may be struggling to fill vacancies, others may have become aware that their workforce is not at all reflective of society, and others still may feel the need to address social inequalities. Consequently, in recent years, increasing numbers of organizations began to turn towards Diversity Management.

Diversity Management is an active process, involving the coordination and direction of employee differences to ensure the meeting of organizational goals (Friday and Friday 2003). That being said, Diversity Management does not refer to controlling or containing it, nor does it involve the assimilation of employees to fit the organization's existing culture (IBEC 2003; Thomas 1990). Instead, diversity management refers to enabling each member of a workforce to perform to their potential (Thomas 1990). There is no one, best, method for, or model of, Diversity Management, yet there are a number of steps an organization can take.

First, organizations can undertake an analysis of their current diversity standing. This involves constructing a profile of the level of diversity that exists among the organizations workforce, and its customer/client base. The purpose of this is to begin assessing whether there is an obvious disparity between the population and the workforce, or the workforce and the customer base it serves. Another element of this analysis concerns the examination of current employment processes to identify areas relating to diversity that require addressing (IBEC 2003), for reasons mentioned earlier in this chapter.

Next, organizations should take a comprehensive approach to establishing the key objectives behind the diversity strategy. Essentially, the organization should set out what they want to achieve, how they plan to do so, and a timeframe (IBEC 2003) for activities and implementation. This vision, and the motives behind the organization's interest in diversity, should be communicated to existing employees, as should the ways in which managing diversity will benefit the organization (Kreitz 2008).

While it may sometimes be appropriate to devise new policies, amendment of existing policies and procedures is also important. Following this planning stage, implementation of initiatives should be rolled out. A core initiative or activity, considered by some to be a 'must' (Lai and Kleiner 2001), and proposed to be both one of the most widely used activities (Carnevale and Stone 1994), and one of the most visible features (Hite and McDonald 2006), under the umbrella of diversity management, is Diversity Training. Diversity training does not refer to any one specific activity. Instead, the term can be used to describe many workplace diversity management interventions (Ferdman and Brody 1996).

There are two primary diversity training approaches which may be used to reinforce each other, specifically, awareness based training and skills based training (Carnevale and Stone 1994). The aim of awareness based training is to heighten awareness of diversity. Objectives of awareness training include providing employees and managers with information about diversity, and heightening awareness of, and sensitivity to, diversity. This is achieved by uncovering assumptions and biases, assessing current attitudes and values, correcting stereotypes and myths, and fostering group and individual sharing of views and experiences.

The aim of skills based diversity training is to provide employees with skills that empower them to effectively deal with diversity in the workplace (Carnevale and Stone 1994). Specific competencies and skills are required if employees are to work successfully as members of a diverse group. Skills based diversity training initiatives can assist in developing those skills and competencies (Moore 1999). As such, skills based training is principally focused on behaviour, and providing tools to promote effective interaction in a diverse organization (Carnevale and Stone 1994).

Successful implementation of training and other Diversity Management activities, policies and processes, requires diversity strategies to be seen not solely as a Human Resource Management (HRM) issue, rather, as vital to the achievement of organizational goals. Once the Diversity Management program has been implemented, it is crucial that it is regularly assessed, and, if found to not be achieving the desired objectives, amended (IBEC 2003; Cox and Blake 1991).

While a useful step in better integrating and managing diverse employees, diversity management can, regardless of its best intentions, be divisive (O'Donovan 2018). Diversity management and diversity training initiatives may sometimes result in attempting to understand employees by "assigning" them to, and viewing them in light of assignment to, a particular grouping. Organizations may, for example, primarily focus on a small number of demographics, whether gender, age or sexual orientation, and only consider these elements of employee diversity. This rigid categorization has the potential to ignore other aspects of an individual's diversity, potentially resulting in individuals feeling marginalized or misunderstood. Consequently, while organizations should build a foundation on diversity management, and employee diversity training, to help enhance understanding and open channels for communication, it is recommended that they do so with the ultimate aim of creating an organizational environment that is inclusive for all employees (O'Donovan 2017; Sabharwal 2014).

Individuals need to feel, and actually be, included in their place of work (Davidson and Ferdman 2002). Diversity initiatives will be more successful if managers engage and use processes that foster equity, consensus and empowerment among, and of, employees (Anderson 1993). Inclusion is the crux of organizational diversity efforts (Sabharwal 2014), and, ultimately, organizations should move towards inclusion.

According to Ferdman (2013: 4):

> Inclusion involves how well organizations and their members fully connect with, engage, and utilize people across all types of differences.

Essentially, inclusion concerns holistically viewing the employee, integrating both their similarities and differences into the fabric of the organization. Employees are allowed to be their full selves while in the workplace, and both differences and similarities are leveraged to improve organizational functioning and performance.

Creating and maintaining inclusion is an ongoing process. A number of contributory factors have been identified which can assist in developing and maintaining employee perceptions of inclusion. Of course, as all employees are different, the factors may not automatically make all employees feel included, or to the same extent, as what constitutes inclusion varies for different people (O'Donovan 2018).

The first contributory factor that organizations can develop, encourage and maintain relates to teams. Having a sense of team has been identified as important for inclusion, therefore, organizations should identify ways in which a team orientation can be created (O'Donovan 2015). While the structure of some organizations may more naturally result in team working, other, more individually structured organizations could schedule team meetings, or encourage a sense of team from the onboarding stage, or during training activities, by grouping individuals from different departments together to allow the development of relationships across the organization (O'Donovan 2018).

The second contributory factor that assists in moving towards an inclusive organization concerns stability. In this context, stability concerns the related concepts of relationships (in the workplace) and familiarity (in a unit or department). This essentially means that individuals should be allowed to form and maintain relationships in the workplace, and ideally not be frequently moved to different teams or functions if possible.

Employee engagement also carries importance for a sense of inclusion. Perceiving respect from colleagues, and being willing, and believing there exists the freedom, to offer opinions on work related matters, are contributory factors. Feeling respected and free to engage in dialogue in the workplace are elements of employee engagement, which suggests a link between engagement and inclusion. Organizations should, in conjunction with inclusionary efforts, also focus on employee engagement, as both engagement and inclusion arguably support and reinforce each other. For some individuals, however, being asked for their opinion directly or publicly can be quite uncomfortable, due to, for example, personality type or cultural norms, leaving them feeling un-included. To increase engagement through participation, it may therefore be useful to allow individuals to give opinions or suggestions privately, perhaps via email, a suggestion box, or private meetings.

The final factor concerns management. If managers are seen to behave inclusively, this symbolises the importance of inclusion to others. Similarly, if management are seen to encourage employees to utilise their differences to assist in their work, this symbolises acceptance and valuing of differences. Managers also have a role to play in reinforcing the culture of the organization, and so have a role to play in reinforcing a culture of inclusion. Furthermore, managers can act as driving forces, or champions, for the other contributory factors identified above. In order for managers to behave inclusively, they must perceive inclusion. This highlights that inclusion must be considered at the individual, team, and organizational level (O'Donovan 2018; O'Donovan 2015).

As identified at the outset, the third primary factor of interest in this chapter is Organizational Citizenship Behaviour (OCB). While OCB may not initially appear to be in any way related to Diversity and Inclusion, their relationship, particularly the manner in which they link to create positive organizational functioning, and impact Knowledge Management and Learning, will be illustrated later in the chapter.

Discussion of concepts building up to OCB has been around for over 80 years in various guises, for example Willingness to Cooperate (Barnard 1938), Supra Role Behaviour (Katz and Kahn 1966) or Citizenship Behaviours (Bateman and Organ 1983). In the 1980s, the construct of OCB was explicitly developed (Markóczy et al. 2009; Becton et al. 2008; Borman 2004). OCB has been defined as:

> Individual behaviour that is discretionary, not directly or explicitly recognised by the formal reward system, and that in the aggregate promotes effective functioning of the organization (Organ 1988: 4).

Clarity has been provided for some terms in this definitions, specifically that:

> By discretionary, we mean that the behaviour is not an enforceable requirement of the role or the job description, that it is the clearly specifiable terms of the person's employment contact with the organization; the behaviour is rather a matter of personal choice, such that its omission is not generally understood as punishable (Organ 1988: 4).

Essentially, OCB refers to behaviours undertaken by employees of their own volition, which are not required as part of their job or task fulfilment, and so are not practicably enforceable by superiors, but often help the organization in some manner (Tambe and Shanker 2014; Vigoda-Gadot 2006; Jahangir et al. 2004). While OCBs are deemed discretionary in nature, superiors:

> Presumably value such behaviour, in part because they make their own jobs easier and free their own time and energy for more substantive tasks (Bateman and Organ 1983: 588).

Indeed, it has been argued that organizations could not survive without employees undertaking in OCBs (Tambe and Shanker 2014; Jahangir et al. 2004).

OCB is a multi-dimensional construct (Markóczy et al. 2009; Vigoda-Gadot 2006; Podsakoff et al. 2000). Comprehensive research undertaken by Podsakoff et al. (2000) noted that 30 potential forms of OCB were identifiable at that time. They discovered, however, that based on those forms, the typical behaviours which constitute OCB can be categorised into seven dimensions, specifically:

1. Helping Behaviour
2. Sportsmanship
3. Organizational Loyalty
4. Organizational Compliance
5. Individual Initiative
6. Civic Virtue

7. Self-Development.

Helping behaviour concerns helping others in the organization with work related problems, or helping them to avoid those problems (Podsakoff et al. 2000). Sportsmanship relates to an individual's willingness to accept work-related inconveniences without complaining (Organ 1990). Sportsmanship also concerns employees maintaining a positive attitude when things do not go their way, and not being offended when others do not follow their suggestions. "Good Sports" in this context, are also willing to sacrifice their personal interests for the good of the group, and do not take rejection of their ideas personally (Podsakoff et al. 2000).

Organizational loyalty refers to the practice of promoting the organization to 'outsiders', and protecting and defending the organization from threats in the external environment. Employees displaying these behaviours also remain committed to the organization when it is operating under difficult conditions, for example the current Covid 19 pandemic, or recessionary times.

The dimension of organizational compliance concerns acceptance of organizational rules, regulations and procedures, and adhering to them, even when adherence is not monitored or observed. This dimension is interesting, as while all employees are expected to always adhere to the organizations rules, some do not. As such employees who do, even when they are not being monitored, are considered to be displaying good citizen behaviours.

Individual Initiative can be categorised as extra-role behaviour as it involves engaging in behaviours that are in fact task-related, but at a level that is so far beyond what is expected (going 'above and beyond') that it begins to take on voluntary characteristics. Employees displaying civic virtue possess a macro-level commitment to the organization. This is indicated by an employee's willingness to actively participate in the organizations governance, monitor the organizations environment for threats and opportunities, and to look out for the organizations best interests, even if at personal cost. Civic virtue also reflects an employee's recognition of being part of a larger unit or whole. The seventh grouping, Self-Development, comprises voluntary behaviours undertaken by employees to improve their knowledge, skills and abilities (Podsakoff et al. 2000).

It is, of course, likely that other types of OCBs exist. Some OCBs may be more customer focused, or may be specifically found to be practiced among employees in particular jobs. Regardless of the type of OCB, as noted above, they present advantages. So too does Diversity, and Inclusion. While this chapter is specifically interested in Knowledge Management and Learning, a brief discussion of advantages associated with Diversity, Inclusion and OCBs is first needed.

2 Advantages Associated with Diversity, Inclusion and OCBs

As noted, Diversity and Inclusion are not the same. There is a well-established Business Case for managing diversity, detailing a set of advantages associated with diversity in the workplace. Many of these advantages are, however, mirrored by the advantages associated with inclusion, although inclusion presents some additional advantages (O'Donovan 2015, 2017). Consequently, the advantages associated with both Diversity and Inclusion will be explored jointly in this chapter. Figure 1 presents these advantages.

The Cost Savings advantage can be broken into three parts, and primarily concerns the negative impact the mismanagement of diversity has on an organization's bottom line. The first part of this argument relates to turnover levels. Turnover among diverse employees is a costly and significant problem, as are the consequential added recruiting, staffing and training costs per person. Some employees who are on the receiving end of discrimination by management or colleagues, whether real or perceived, will leave the organization. Other employees who feel their needs as individuals are not being met by the organization, or who feel they are not being reasonably facilitated in requests, will leave. It is also likely that some employees who feel 'out of the loop' will leave. In a related vein, the second component of the cost savings argument concerns lowering absenteeism rates. Absenteeism can amount to significant costs, and can occur when individuals do not feel secure about their status, as such insecurity prevents employees from fully engaging at work. If employees do not feel comfortable or feel they belong in the workplace, they are more likely to not attend work (or to leave). The final consideration under cost savings centres on lawsuits based on, for example, sexual, age, or race discrimination, or, more specifically, a strategic organizational effort to avoid them (Robinson and Dechant 1997; Marchington and Wilkinson 2005). Organizations can begin their efforts to lower legal costs by using the aforementioned

diversity training events to train employees in what constitutes acceptable and unacceptable behaviour. Enforcement and support of these norms will help to make it clear that discrimination is not accepted, which should ideally see a reduction in law suits founded on discrimination claims (O'Donovan 2015).

The next advantage, winning the competition for talent, or the "talent war", refers to the attraction, retention and promotion of employees from different demographic groups (Robinson and Dechant 1997; Lockwood 2005). The future and sustainability of an organization is dependent on the quality of talent it attracts and retains (Gardenswartz and Rowe 1998); employees do the work that allows the organization to operate. Organizations that recruit, develop, retain, and promote diverse employees are more likely to have an edge over competitors, as other talented candidates will be attracted to an organization that values their capabilities. These employees will also be more willing to invest in productive activity, and less likely to leave, if they believe they are being treated fairly, and that career opportunities are available to them (Robinson and Dechant 1997). Effective retention is a foundation stone for

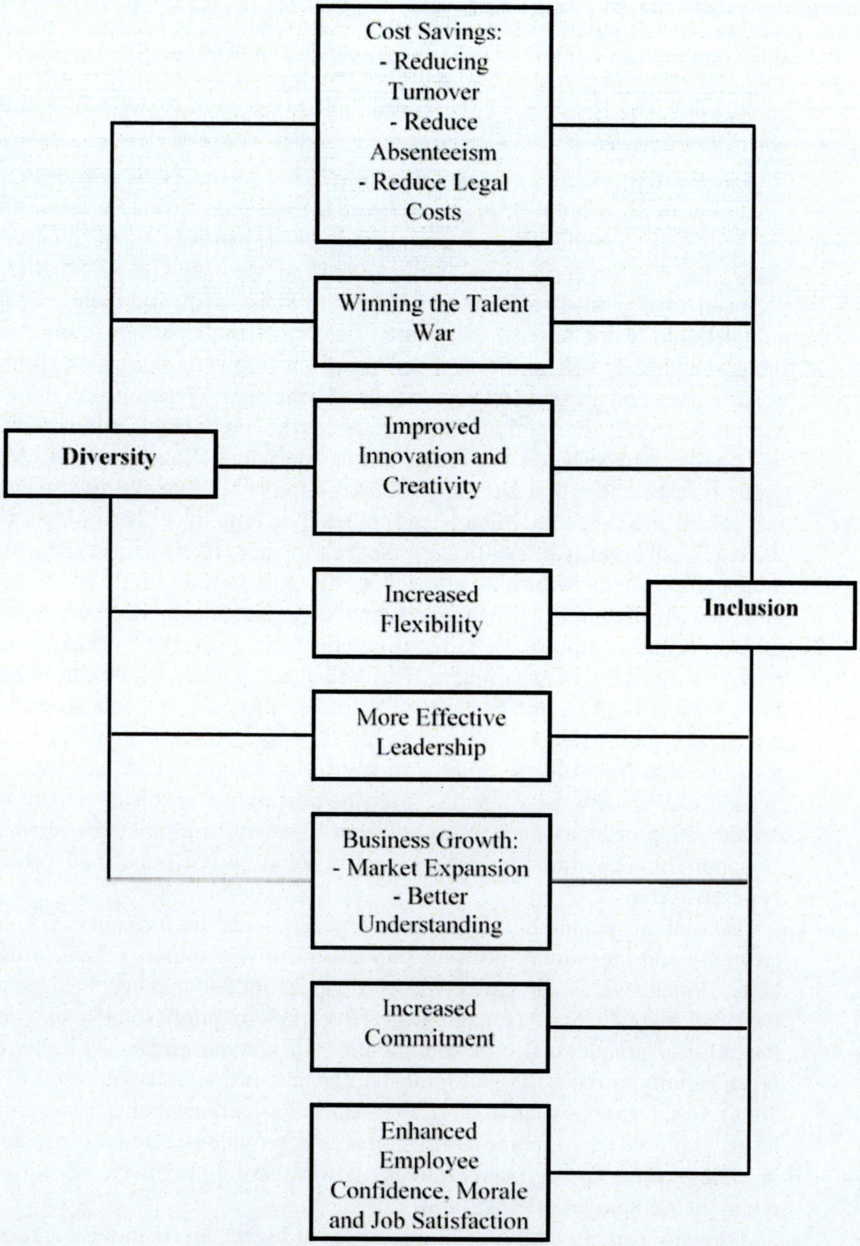

Fig. 1 Advantages Associated with Diversity and Inclusion. *Source* Author (O'Donovan (O'Donovan 2015))

organizations attempting to win the talent war. Inclusion of diverse employees can aid in retention, as commitment to diversity indicates to employees that the organization cares for them as individuals (Espinoza 2007). If organizations do not effectively manage diversity issues, diverse talent will leave in favour of a competitor who does (Bagshaw 2004).

Another opportunity arising from workforce diversity is greater employee creativity and innovation (Monks 2007; van Knippenberg 2007; European Commission 2003). In an inclusive organization, individuals can use their personal resources to do what they do best, meaning employees can use their diverse experiences and perspectives (Davidson and Ferdman 2002) to assist them, and their colleagues, in completion of work activities. Essentially, as diversity shapes how individuals view things, employees will have different ways of viewing or thinking about problems or opportunities and the performance of tasks (Flynn 1995; Waters 1992). If the varying approaches, views or opinions are considered, the likely result is that management will be able to make better and more informed decisions (Bagshaw 2004). Managing diversity, especially through an inclusionary approach, can make diverse employees feel valued and supported, which tends to result in employees becoming more innovative (Eisenberger et al. 1990), as employees are more likely to take risks, or try new things, if they believe they are allowed, or have the support, to do so. Diversity can also increase the quality of team problem-solving, as, again, diversity among team members enables employees to see problems from an array of perspectives, based on their wide range of experiences. This will again, potentially, produce better decisions (Kim 2006; Cox and Blake 1991; Lowther 2006). Less emphasis on employee conformity should also improve creativity (Lowther 2006). If individuals approach the same task from diverse points of view, task-related conflicts are more likely to occur. Dealing with these conflicts has the potential to result in a more thorough consideration of all aspects and approaches to solving a problem or capitalising on an opportunity, culminating in more creative solutions (Griffin and Moorhead 2006; Justesen 2001).

As well as potentially improving organizational performance via enhanced creativity and innovation, diversity can also positively impact organizational flexibility. If inclusively managing diversity, organizations should become less standardised, and more fluid, as managing diversity likely requires some adaptation of, or flexibility in, practices. This fluidity should, in turn, create greater flexibility, enabling organizations to react to environmental changes faster, and at less cost (Lowther 2006). Organizations used to offering flexible work arrangements, for example, may be better placed to overcome skills shortages or provide alternatives to redundancies in difficult times through career breaks or job-sharing initiatives, allowing them to retain their employees (Kim 2006).

Diversity can also improve effectiveness at higher levels in the organization. A diverse pool of individuals at top management levels can prevent a myopic perspective, for the same reasons supporting the argument of enhanced creativity and innovation. Diverse leaders, who reflect the diversity of the organization, are better able to understand the rest of the organization's employees, in addition to sending a signal that diverse employees have the same opportunity for advancement. This can result in yet further diversity at higher levels, and potentially assist in retention. Increased awareness, appreciation and understanding of diversity can also help management become more effective in cross-cultural business situations (van Knippenberg 2007; Flynn 1995). If nothing else, at a basic level, good diversity skills are compatible with good people management skills, and so focussing on management's ability to supervise a diverse workforce can result in improvement of their overall people management skills (Cox and Blake 1991).

Driving Business Growth, as an advantage associated with diversity, concerns organizations managing diversity to leverage possible opportunities. First, organizations can use workforce diversity to gain an increased understanding of the marketplace in which they operate (van Knippenberg 2007; Cox and Blake 1991). Improved access to new market segments and improving performance in existing markets have been found to be benefits of diversity (Centre for Strategy and Evaluation Service 2003). As the marketplace, customers and suppliers are diverse (Lowther 2006; Farrer 2004), it makes sense that the understanding needed to market to diverse demographics, and to respond to their needs, more naturally resides in employees with the same background (Flynn 1995; Lowther 2006). Similarly, a diverse and inclusive workforce can assist in ensuring that the products and services developed and offered by the organization are respectful and mindful of customers or customer's clients, and better match their needs. The reason for this is that if the workforce is reflective of the organization's customer base, and is willing and able to use that similarity to improve product or service offerings, the organization's outputs can be closer aligned with customer desires. This can result in increased customer satisfaction (Ferdman 2003; Doke and Beagrie 2003), which carries the potential for repeat sales, improving the bottom line and market share. In addition, diversity provides a good image to an organization's customer base, which can enhance company branding (Espinoza 2007) efforts and consumer loyalty to the organization.

Additionally, employees who believe their employer supports them have a tendency to be more productive. This increase in productivity positively impacts the organization's bottom line (Carnevale and Stone 1994; Espinoza 2007). Again as discussed earlier, commitment to diversity, particularly an inclusive one, enables every employee to contribute their individual ideas, talents, and skills to the organization, and this ultimately drives the organization's bottom line (Eisenberger et al. 1990).

Inclusive organizations may also benefit from employee willingness to assist in global challenges, including varying laws and regulations, language barriers and cultural barriers (Doke and Beagrie 2003). Employees in an inclusive organization may, for example, be willing to openly discuss their national culture, thereby assisting the organization in better understanding, and adapting to, that culture, whether for the purposes of setting up an operation in that area or tapping a potential market. Moreover, inclusion can result in improved productivity and less errors as employees will likely feel they can give opinions or speak frankly, which can result in the production of products of a higher quality (Doke and Beagrie 2003). Less errors may also positively impact the organizations bottom line, as less profit will be lost to wastage.

Inclusion can also result in increased knowledge transfer, group cohesion and a more positive group climate, thus, a better work environment (Doke and Beagrie 2003). Increased knowledge transfer may pave the way for smoother team, departmental and cross departmental functioning, enhancing business operations. Improved group cohesion and a positive group climate will likely assist in the creation of a team culture, which can help to improve overall group performance, again contributing to business growth and bottom line improvements.

Perceived inclusion can also increase commitment from employees (O'Donovan 2015; Doke and Beagrie 2003), whether to the job or organization, and enhance productivity, which is likely to enhance performance (O'Donovan 2015). Increased commitment has the potential to see reduced absenteeism and turnover levels, allowing for stability in the workplace, and, in line with earlier discussion, cost savings through reduced recruitment and selection costs. Inclusion can enhance employee confidence (O'Donovan 2015; Doke and Beagrie 2003). Enhancing confidence may improve employee's perceptions of their ability, which can positively impact performance. In addition to enhancing performance, increased confidence and commitment may increase the likelihood of employees going beyond their required tasks and undertaking OCBs. Furthermore, inclusion can also result in enhanced morale, feelings of support and job satisfaction which have been cited in literature as antecedents of OCBs. If inclusion results in these factors, it is arguably evident that a relationship exists between inclusion and OCBs (O'Donovan 2015).

OCBs, as indicated earlier in the chapter, are, by their very nature, advantageous for the organization. Not every instance of OCB by an employee contributes to positive organizational outcomes (Organ 1997), however, there are a number of reasons why it can be expected that OCBs will contribute to the effectiveness of the organization. OCBs have, for example, the potential to enhance the productivity of colleagues and supervisors or mangers, and can help coordinate organizational activities. OCBs can also increase performance levels and ensure performance stability, and some behaviours can assist in attraction and retention of employees (Borman 2004). OCBs may also increase available organizational resources, and decrease the need for some formal or expensive control methods (Podsakoff and MacKenzie 1997).

The description of the seven broad categories of OCBs presented earlier points to the potential advantages associated with OCBs, however, it is worthwhile to consider the possible advantages more fully. As noted, helping behaviour concerns helping others with work related problems, or helping them to avoid them. If employees are assisting other employees, this frees up management time for other activities, enhancing managerial efficiency. Helping colleagues in the workplace can also help to create and maintain a feeling of "togetherness" or team spirit, which is a contributory factor for perceived inclusion, engagement and commitment. Improvement in these areas carries the potential to assist in retention of employees, saving on recruitment costs. Furthermore, helping colleagues can take the form of ad hoc, on the job training, possibly reducing training costs, presuming, of course, that employees are passing on correct information. Helping others to avoid work-related problems can assist in reducing mistakes, which can be beneficial for organizational efficiency, and assist in reducing wastage costs.

Employees displaying Sportsmanship behaviours by accepting work-related inconveniences with minimal complaining can assist in maintaining good morale in the workplace. Good morale can help to maintain good levels of employee engagement, which can again enhance commitment and retention. Employees who are willing to sacrifice their personal interests for the good of the team or the organization are also more likely to make decisions that benefit their team as a whole, or the organization, rather than simply serving their own interests, resulting in both ethical and performance oriented decisions.

The set of behaviours termed Organizational Loyalty can assist in recruitment efforts, as employees undertaking these OCBs are likely to promote the organization to those outside it, making it appear to be a good place to work. As mentioned earlier, employees displaying these behaviours remain committed to the organization when it is operating under difficult conditions. This is advantageous for the organization as it may essentially give it time; time to address its adverse circumstances while retaining many of its employees. Employees may, for example, be willing to temporarily accept wage freezes or reduced shifts until the organization has resumed usual activities.

If employees are undertaking behaviours under the Organizational Compliance category of OCBs, they are likely to be following the rules and regulations of the organization even when not being monitored or observed. One advantage associated with this is that the more employees who are adhering to organizational 'rules', the less formal, continual, observation and monitoring of employees is required, freeing up supervisory and managerial personnel to undertake more strategic duties. This is not to say that observation of behaviour for the purposes of Performance Management is not necessary, rather micro-management is not. Moreover, if employees follow 'rules' in the form of policies and procedures, they may be less likely to make costly mistakes, or interact with other colleagues in a manner which may result in claims of discrimination or harassment.

As explained earlier in this chapter, Individual Initiative OCBs relate to going 'above and beyond' while undertaking task related behaviours to the extent that the task performance begins to take on voluntary characteristics. This can see employees exceeding requirements and striving to work to their full potential, improving outputs or organizational functioning. The presence of committed, high performing team members can assist in creating a high performance work culture. Such cultures should be monitored, however, to ensure employees are not over-exerting themselves in a manner that could result in exhaustion.

Employees displaying Civic Virtue behaviours can assist in the day to day functioning of the organization by identifying opportunities, informing management of issues and working with others to resolve them. By joining committees and task forces, employees can act as a communication conduit between management and employees, or between different teams, spreading relevant information in a timely manner.

Finally, as will be further discussed later in the chapter, when employees engage in Self-Development by voluntarily trying to improve their knowledge, skills and abilities, they are potentially enhancing not just their own skill set, but the skills available to the organization. Voluntarily improving knowledge and skills can also, if task related, lower training costs, and result in ad hoc training of other employees.

In addition to the advantages overviewed in this section, Diversity, Inclusion and OCBs arguably have a role to play in learning, and in organizational Knowledge Management efforts. The final section of this chapter will discuss that role and relationship.

3 Learning and Knowledge Management: The Impact and Role of Diversity, Inclusion and OCBs

Knowledge Management refers to the strategic and systematic effort made by organizations to define, capture, retain, and share the knowledge and experience of employees (King 2005). Knowledge Management is of crucial importance for organizational performance, both in the present, and in terms of ensuring future operations. First, capturing a broad range of experiences, ways of viewing threats and opportunities, and knowledge possessed by employees can improve decision making. Second, an organization, without employees, arguably has no knowledge; people possess the knowledge and skills that result in tasks being undertaken and completed, allowing the organization to perform and achieve its goals. Third, as employees possess the necessary knowledge and skills required for organizational functioning, when they leave the organization, they take that knowledge and skill set with them. Without systematic efforts to capture, store, and transmit that knowledge to other employees, organizational effectiveness and future performance ability will likely be negatively impacted. Again, if, for example, a significant number of the higher performing employees were to leave an organization, the knowledge and skills that (in conjunction with work ethic and effort) make them a high performing employee leaves with them.

In comparison, learning is a complicated process, and is difficult to define (Kelly 2002). A primary reason for this difficulty is that learning is an internal process. In addition, it is accepted that children and adults learn in different ways. It is also accepted that, as adults, different people learn differently. Some learn best by doing (kinesthetic learning) i.e. by trying an activity and adjusting performance until they undertake that activity well. Others are visual and learn by observation i.e. watching another person undertake a task. Others learn by undertaking formal, classroom type training or by listening (auditory learning), while others learn best having read written instruction.

As discussed at the outset, the purpose of this chapter is to consider and discuss the potential role, and impact, of Diversity, Inclusion and OCBs in Learning and Knowledge Management. Table 1 presents an initial overview of this discussion and hypotheses.

As noted earlier, people are different in a vast number and manner of ways. Consequentially, learners are diverse. This means that in order to best facilitate learning

Table 1 How Diversity, Inclusion and OCBs Likely Impact Learning and Knowledge Management

Factor	Learning	Knowledge Management
Diversity	• Employees learn differently • Training approaches and formats should be varied • Resources should be made available in different formats • Learning, or retention of information, should be tested differently	• Employees who possess knowledge have different communication styles • Different cultural expectations may exist regarding sharing knowledge • Different perceptions exist regarding appropriateness of asking for information
Inclusion	When inclusion is perceived, employees are: • More likely to be committed to learning • More engaged in the learning process • More likely to engage in self-directed learning • More likely to ask for help with learning activities • More likely to help others learn	When inclusion is perceived, employees are: • More likely to share knowledge and expertise with others in the organization • More likely to engage in strategic knowledge management efforts • More likely to stay in the organization, enabling the retention of knowledge
OCBS:		
Helping Behaviour:	May result in attempting to assist others to learn in a way that suits the individual	More likely to share knowledge and expertise with others in the organization
Sportsmanship:	More likely to be understanding of the fact that others have a different way of learning, so willing to alter training approaches, or engage in learning events not tailored just to their learning style	Likely to continue to share ideas and knowledge, even if ideas are not always acted upon, or expertise sharing reciprocated
Organizational Loyalty:	As part of efforts to protect the organization, Organizational Loyalty behaviours may include upskilling to improve performance, even when not required to do so by the organization, or when attendance at external training for the purposes of learning cannot be paid for by the organization	One way in which the organization can be defended from threats in the external environment is by the sharing of pertinent external information by employees when they become aware of it

(continued)

Table 1 (continued)

Factor	Learning	Knowledge Management
Organizational Compliance:	Learning by not just undertaking, but fully engaging in, mandated training that may not be directly linked to task requirements or match individual learning styles, arguably falls under the category of compliance	One element of a Knowledge Management Strategy may be to capture knowledge from existing high performers in order to assist in the development of others. Compliant behaviour oriented employees may be more likely to engage in job analysis and performance appraisal activities to assist in establishing the knowledge and skills a "good" performer possesses, which can be stored and utilized by the organization
Individual Initiative:	Employees possessing and demonstrating Individual Initiative behaviors could be considered to be more likely to strive to learn as much about their tasks as possible in order to improve performance	Individuals displaying these behaviours, who strive to undertake their tasks as well as possible, should be included in Knowledge Management efforts to elicit their task related knowledge and expertise
Civic Virtue:	Commitment to the organization under this categorization should result in employees understanding the value of employee learning in the organization, and so result in increased commitment to learning and engagement in learning activities	Employees displaying Civic Virtue behaviors recognize they are part of a larger whole, and so should be willing to engage in strategic knowledge management activities, recognizing their role in organizational success
Self-Development:	This entire categorization is directly linked to both learning and knowledge management, as it concerns the voluntary undertaking of activities by employees to improve their knowledge, skills and abilities. Continual learning resulting in improvement of skill sets and knowledge by any cohort of employees, particularly if Knowledge Management strategies are in place, also improves the organization's skill set, as the organizational skill set is an amalgamation of individual employee skill sets	

Source Author (2021)

in the origination, diversity in training approaches should be implemented to match diversity of learning styles among learners. This may include providing different types of resources that satisfy the same outcome, or running a number of events that address the same training need. In addition to a classroom type training session, for example, the organization could transmit a recording of the session that employees who need to can refer back to. A transcript of the recording could also be developed for employees who learn best by reading. Where possible, building in application of content via examples and role plays will assist more 'hands-on' learners. Moreover, it should be remembered that employees process information and learn at different speeds. It may take some learners longer than others to be able to put learning into practice. Employees will also have different retention abilities based on their learning style, and so may find it difficult to recall skills and knowledge that they do not often need to access. As such, refresher training for important, but not often used information, should be a priority.

There are other very practical concerns associated with diversity and facilitation of learning. Resources for training events should be prepared with differing requirements in mind. Variations of resources should be developed, for example, which cater for employees with sight or hearing difficulties. If employees are not presented with accessible materials, and an even playing field, for learning they are likely to feel excluded from learning events, which for many, will make the internal process of learning more difficult.

Regarding Knowledge Management, diversity reminds us that employees have differing communication styles, and so will have different ways of communicating what they know. It does not automatically follow that a high performing, knowledgeable employee will have the ability to transmit their knowledge to others. As such, making sure to systematically catalogue that employee's knowledge to make it suitable for transmission to others is important. Additionally, based on cultural theories, it is arguable that cultural complexities may give employees pause when deciding whether they should share knowledge with, or seek knowledge from, others. Employees from High Power Distance (Hofstede 1983) cultures, for example, where individuals are expected to defer to authority, and superiors are not expected to directly comment on the work of subordinates, may be reluctant to share a different take on a problem or opportunity in meetings for fear of appearing to overstep. Superiors may feel it is not their place to share knowledge with employees under their span of control, instead believing that should come under the remit of HRM controlled Learning, Training and Development activities. In contrast, employees from Collectivist cultures, where group loyalty and harmony is valued (Hofstede 1983; Hofstede 1999), may find the concept of freely sharing knowledge and helping others to develop skills a much more natural activity. Furthermore, employees form Individualistic cultures (where individual achievement is valued), or high Uncertainty Avoidance cultures (where the unknown is feared (Hofstede 1983)), may be reluctant to share or seek knowledge from others in the organization. In a similar vein, while not true of all, some Type A personalities (Friedman and Booth-Kewley 1987), who can have a desire to be in control, may not want to relinquish information that allows them to be high-achieving, for fear of relinquishing some control or no longer being perceived as 'the best'. Those employees could be encouraged to contribute to Knowledge Management efforts by being invited to contribute their knowledge in high profile ways, for example involvement in developing training or induction tools, participation in Think Tank teams, or presentations at meetings.

Considering diversity reminds us that people learn, process, transmit and view ownership of knowledge differently. Organizations should also remember that perceived inclusion may impact the extent to which employees commit to, and engage

also improves the overall organizational skill set, as the organizational skill set is an amalgamation of individual employee skill sets.

4 Conclusion

Organizational performance is a function of individual performance; without the work of employees, nothing would be achieved. As such, organizational success is dependent on proper utilization of employee knowledge and skill sets, both in terms of employees using their knowledge and skills to perform effectively, and also in terms of organizations encouraging learning (and development), and making efforts to collect, store and transmit employee possessed knowledge for future use. This chapter has proposed that effectively doing the latter requires consideration of three factors, specifically Diversity, Inclusion, and Organizational Citizenship Behaviour. At the core of this chapter are three arguments. First, Diversity impacts how individuals learn, therefore, organizational activities developed with a view to encouraging learning should be done so to reflect diversity. Second, an inclusive approach to diversity is more likely to result in employees being engaged in both Learning and Knowledge Management activities, and in undertaking OCBs. Third, it is argued categories of OCBs are likely related to willingness to learn and willingness to assist in organizational Knowledge Management efforts. Ultimately, organizations are therefore encouraged to reflect employee diversity in Learning and Knowledge Management events, and to develop an inclusive organization to increase the instances of OCBs, and enhance employee commitment to, and engagement with, Learning and Knowledge Management.

References

Anderson, J. A. (1993). Thinking about diversity. *Training and Development, 47*(4), 59–60.

Bagley, C., & Abuaker, M. (2017). Muslim woman seeking work: An English case study with a dutch comparison, of discrimination and achievement. *Social Sciences, 6*(1), 17.

Bagshaw, M. (2004). Is diversity divisive? A positive training approach. *Industrial and Commercial Training, 36*(4), 153–157.

Barnard, C. I. (1938). *The functions of the executive*. Cambridge, MA: Harvard University Press.

Bateman, T. S., & Organ, D. W. (1983). Job satisfaction and the good soldier: The relationship between affect and employee "Citizenship". *Academy of Management Journal, 26*(4), 587–595.

Becton, J. B., Giles, W. F., & Schraeder, M. (2008). Evaluating and rewarding OCBs: Potential consequences of formally incorporating organizational citizenship behaviour in performance appraisal and reward systems. *Employee Relations, 30*(5), 494–514.

Borman, W. C. (2004). The concept of organizational citizenship. *Current Directions in Psychological Science, 13*(6), 238–241.

Carnevale, A. P., & Stone, S. C. (1994). Diversity: Beyond the golden rule. *Training and Development, 48*(10), 22–39.

Centre for Strategy and Evaluation Service. (2003). *The costs and benefits of diversity: Executive summary*. United Kingdom: Centre for Strategy and Evaluation Services.

Commission, European. (2003). *Employment & Social Affairs. The costs and benefits of diversity*. United Kingdom: CSES.

Cox, T. H., & Blake, S. (1991). Managing cultural diversity: Implications for organizational competitiveness. *Academy of Management Executive, 5*(2), 34–47.

Davidson, M. N., & Ferdman, B. M. (2002). Inclusion: What can I and my organization do about it? *The Industrial-Organizational Psychologist, 39*(4), 80–85.

Doke, D., & Beagrie, S. (2003). How to…Manage a diversity programme. *Personnel Today, 2*(4), 31–34.

Eisenberger, R., Fasolo, P. M., & Davis-LaMastro, V. (1990). Effects of perceived organizational support on employee diligence, innovation and commitment. *Journal of Applied Psychology, 53*, 51–59.

Espinoza, M. (2007). Turning diversity into a competitive advantage. *Financial Executive, 23*(3), 43–46.

Farrer, J. (2004). A Practical approach to diversity. *Industrial and Commercial Training, 36*(4), 175–177.

Ferdman, B. M. (2003). Accounts of inclusion (and exclusion). *The Industrial-Organizational Psychologist, 40*(4), 81–86.

Ferdman, B. M. (2013). Frameworks for Understanding Inclusion. In B. M. Ferdman & B. R. Deane (Eds.), *Diversity at work: The practice of inclusion*. Hoboken, NJ: Jossey-Bass.

Ferdman, B. M., & Brody, S. E. (1996). Models of diversity training. In D. Landis & R. S. Bhagat (Eds.), *Handbook of intercultural training* (2nd ed.). Thousand Oaks, CA: Sage.

Flynn, G. (1995). Do you have the right approach to diversity? Avon, Bank of America and Gannett's models of diversity. *Personnel Journal, 74*, 68–70.

Friday, E., & Friday, S. W. (2003). Managing diversity using a strategic planned change approach. *Journal of Management Development, 22*(10), 863–880.

Friedman, H. S., & Booth-Kewley, S. (1987). Personality, type a behavior, and coronary heart disease: The role of emotional expression. *Journal of Personality and Social Psychology, 53*(4), 783–792.

Gardenswartz, L., & Rowe, A. (1998). Why diversity matters. *HR Focus, 75*(7), 51–54.

Gradín, C. (2018). Occupational segregation by race in South Africa after apartheid. *Review of Development Economics, 23*(2), 553–576.

Griffin, R., & Moorhead, G. (2006). *Fundamentals of organizational behaviour: Managing people and organizations*. Boston: Houghton Mifflin.

Hite, L. M., & McDonald, K. S. (2006). Diversity training pitfalls and possibilities: An exploration of small and mid-size US organizations. *Human Resource Development International, 9*(3), 365–377.

Hofstede, G. (1983). National cultures in four dimensions. *International Studies of Management & Organization, 13*(1–2), 46–74.

Hofstede, G. (1999). Think locally, act globally: cultural constraints in personnel management. *Management International Review, 38*(2), 7–26.

IBEC. (2003). *Ireland's changing workforce: Harnessing diversity in the workplace*. Dublin: Irish Business and Employers Confederation.

Jahangir, N., Akbar, M. M., & Haq, M. (2004). Organizational citizenship behaviour: Its nature and antecedents. *BRAC University Journal, 1*(2), 75–85.

Joplin, J. R. W., & Daus, C. S. (1997). Challenges of leading a diverse workforce. *Academy of Management Executive, 11*(3), 32–47.

Justesen, S. L. (2001), *A study of the dynamics inherent in the relation-ship between innovation and diversity*, Master's Thesis, Denmark: Copenhagen Business School, cited in Bechtoldt, M. N., De Dreu, C. K. W., Nijstad, A. (2007), Team personality diversity, group creativity, and innovativeness in organizational teams. Sustainable Diversity EURODIV Working Paper No. 1. 2007 [available online] http://www.susdiv.org/uploadfiles/SD2007-001.pdf

Katz, D., & Kahn, R. L. (1966). *The social psychology of organizations.* New York: Wiley.
Kelly, L. (2002), "What is learning… and why do museums need to do something about it?", Paper presented at the *Why Learning? Seminar*, Australian Museum/University of Technology Sydney, 22 November.
Kim, B. Y. (2006). Managing workforce diversity: Developing a learning organization. *Journal of Human Resources in Hospitality and Tourism, 5*(2), 69–90.
King, D. (2005). Humanitarian Knowledge Management. In B. Carle & B. Van de Walle (Eds.), *Proceedings of the Second International ISCRAM Conference*, Brussels, Belgium (April)
Kreitz, P. A. (2008). Best practices for managing organizational diversity. *The Journal of Academic Librarianship, 34*(2), 101–120.
Lai, Y., & Kleiner, B. K. (2001). How to Conduct Diversity Training Effectively. *Equal Opportunities International, 20*(5/6/7), 14–18.
Lockwood, N. R. (2005). Workplace diversity: Leveraging the power of difference for competitive advantage. *HR Magazine, 50*(6), 1–10.
Lowther, R. (2006). Embracing and managing diversity at Dell: Introducing flexible working and a women's network to help retain key employees. *Strategic HR Review, 5*(6), 16–19.
Marchington, M., & Wilkinson, A. (2005). *Human resource management at work: People management and development* (3rd ed.). London: CIPD.
Markóczy, L., Vora, D., & Xin, K. (2009). Forbearance in organizational citizenship behaviour. *The International Journal of Human Resource Management, 20*(2), 321–347.
Monks, K. (2007). *The business impact of equality and diversity: The international evidence.* Dublin: The Equality Authority and National Centre for Partnership & Performance.
Moore, S. (1999). Understanding and managing diversity among groups at work: Key issues for organizational training and development. *Journal of European Industrial Training, 23*(4/5), 208–217.
Morley, M., Moore, S., Heraty, N., Linehan, M., & MacCurtain, S. (2004). *Principles of organizational behaviour: An Irish Text* (2nd ed.). Dublin: Gill & Mac-millan.
Mosca, I., & Wright, R. E. (2018), Effect of retirement on cognition: Evidence From the Irish Marriage Bar. *Demography, 55*, 1317:1341.
O'Donovan, D. (2015), *Diversity, inclusion and organizational citizenship behaviours: A study of nurses in the Irish healthcare sector*, Unpublished PhD Thesis, Cork Institute of Technology.
O'Donovan, D. (2017), "Inclusion: Diversity management 2.0", In C. Machado & J. P. Davim (Eds.), *Managing organizational diversity: Trends and challenges in management and engineering.* Germany: Springer International Publishing.
O'Donovan, D. (2018). Diversity and inclusion in the workplace. In C. Machado & P. Davim (Eds.), *Organizational behaviour and human resource management: A guide to a specialized MBA course.* Germany: Springer International Publishing.
O'Donovan, D. (2019). HRM in the organization: An overview. In C. Machado & J. P. Davim (Eds.), *Management science.* Germany: Springer International Publishing.
Organ, D. W. (1988). *Organizational citizenship behavior—The good soldier syndrome* (1st ed.). Lexington, Massachusetts/Toronto: D.C. Heath and Company.
Organ, D. W. (1990). The subtle significance of job satisfaction. *Clinical Laboratory Management Review, 4*, 94–98.
Organ, D. W. (1997). Organizational citizenship behaviour: It's construct clean-up time. *Human Performance, 10*(2), 85–97.
Podsakoff, P. M., & MacKenzie, S. B. (1997). Impact of organizational citizenship behaviour on organizational performance: A review and suggestion for future research. *Human Performance, 10*(2), 133–151.
Podsakoff, P. M., MacKenzie, S. B., Paine, J. B., & Bachrach, D. G. (2000). Organizational citizenship behaviours: A critical review of the theoretical and empirical literature and suggestions for future research. *Journal of Management, 26*(3), 513–563.
Redmond, J., & Harford, J. (2010). 'One man one job': the marriage ban and the employment of women teachers in Irish primary schools. *Paedagogica Historica, 46*(5), 639–654.

Robinson, G., & Dechant, K. (1997). Building a business case for diversity. *Academy of Management Executive, 11*(3), 21–31.

Sabharwal, M. (2014). Is diversity management sufficient? Organizational inclusion to further performance. *Public Personnel Management, 43*(2), 197–217.

Serrat, O. (2010). *Glossary of knowledge management*. Washington, D.C.: Asian Development Bank.

Stratton, J. (2008). A Jew singing like a Black Woman in Australia: Race, Renee Geyer, and Marcia Hines. *Journal of Popular Music Studies, 20*(2), 166–193.

Tambe, S., & Shanker, M. (2014). A Study of Organizational Citizenship Behaviour (OCB) and its dimensions: A literature review. *International Research Journal of Business and Management, 1*, 67–73.

Thomas, R. R., Jr. (1990). From affirmative action to affirming diversity. *Harvard Business Review, 68*(2), 107–117.

van Knippenberg, D. L. (2007). *"Understanding diversity", Inaugural address series: Research in management*. Roterdam: Erasmus Research In-stitute of Management.

Vigoda-Gadot, E. (2006). Redrawing the boundaries of OCB? An empirical examination of compulsory extra-role behavior in the workplace. *Journal of Business and Psychology, 21*(3), 377–405.

Waters, H., Jr. (1992). Minority leadership problems. *Journal of Education for Business, 68*(1), 15–20.

Knowledge Codification

Michael O'Meara and Felicity Kelliher

Abstract This chapter offers insight into how knowledge codification has the potential to capture varying levels of tacit content in pursuit of enhanced innovation and ultimately, competitive advantage. It recognizes the need for and importance of codifying knowledge in a variety of ways to allow codified material to move from lower to higher explicit content. We propose that newly articulated knowledge is codified immediately and that informal and formal codification exercises create boundary objects that can be made more complete or explicit through critique, review and reflection on these codified materials. This approach also enhances employee engagement in the codification/re-codification process. The value of involving operators in codification, using existing codebooks, optimizes articulation and codification by inviting a focused review of existing codified material and the subsequent standardization of codified knowledge: it also releases new tacit knowledge within a forum designed to immediately codify this newly articulated knowledge. The proposed knowledge codification framework offers users the potential to apply the principles exhibited in this chapter in practice.

1 Introduction

Our goal in this book is to consider how an organization can manage knowledge in such a way as to facilitate individual and collective learning whilst transforming itself in a continuous way. Earlier chapters have established that knowledge management is understood as the process of creating, sharing and managing the information and knowledge of an organization. Those that have studied codification (Cohendet and Steinmueller 2000; Echajari and Thomas 2015; Gavrilova et al. 2015) have all called for further studies into the nuances of this stage in the knowledge creation process. Thus, in this chapter, we focus on the process of harnessing and amplifying knowledge created within the individual and linking this newly codified knowledge

M. O'Meara · F. Kelliher (✉)
School of Business, Waterford Institute of Technology, Waterford, Ireland
e-mail: fkelliher@wit.ie

C. Machado and J. P. Davim (eds.), *Knowledge Management and Learning Organizations*,
https://doi.org/10.1007/978-3-030-71079-8_2

to an organization's existing knowledge system (Nonaka and von Krogh 2009) in order to facilitate new learning across the organization.

We make a number of assumptions in this assertion that knowledge can be codified. We believe that knowledge is a resource and that individuals in the organization hold knowledge as yet unknown to the wider organization. These individuals know more than they can tell (Polanyi 1966), a reality often referred to as tacit knowledge in the literature. Pursuit of such knowledge requires conscious efforts on the part of the organization to facilitate individual articulation (O'Meara and Kelliher 2017, 2020). While the articulation of this tacit knowledge into a more comprehensible explicit form is a vital first step in harnessing organizational knowledge, this is not the only step in the process. Once knowledge is articulated, tacit knowledge is explicated but, unless this information is captured and disseminated, it remains ineffectual in the wider organizational knowledge system, thereby stifling learning, productivity and innovation (Huber 1991; Romme and Dillen 1997; Dalkir 2005).

Thus, while we believe that knowledge is held by and can be articulated by willing individuals, it is the organization that must ensure that, once articulated, this knowledge is captured in a manner that makes it accessible (Albino et al. 2001). Furthermore, this new knowledge must be applicable to ensure its practical value within different occasions and contexts (O'Meara and Kelliher 2020). In order for newly articulated tacit knowledge to become useable, a codification process should be used to capture it in ways that may be understood, stored and used by all in the organization. Codification captures articulated tacit knowledge making it tangible and mobile enabling the organization to reconstitute it in different places and times by various individuals to support and enhance strategic innovation and competitive advantage (Cowan and Foray 1997; Cohendet and Steinmueller 2000; Cohendet and Meyer-Krahmer 2001). This may require several cycles of recodification before complete, thus this stage in the knowledge creation process warrants attention.

Assuming knowledge is a valuable resource in the pursuit of competitive advantage, this chapter offers insights into how knowledge codification processes can be embedded in an organization in pursuit of competitive goals. Within this chapter, we ascertain that articulated tacit knowledge is of greatest use when codified into explicit forms, which are meaningful and accessible to others. We further debate that codification is a pre-requisite in efficient knowledge transfer and constitutes a key element in the creation of new knowledge (Cowan et al. 2000; Zollo and Winter 2002), and in the pursuit of competitive advantage.

2 Knowledge as a Resource in Pursuit of Competitive Advantage

The interest in the use of tacit and codified knowledge was popularised through the work of Ikujiro Nonaka (1991, 1994) who suggested that, in order for knowledge to be transferred and shared it must be converted into an explicit codified format. This research prompted researchers, including Spender (1996), Eisenhardt and Santos (2000) and Pathirage et al. (2007), to investigate the relationship between the resource-based view (RBV) perspective and the role of knowledge in creating competitive advantage. This informed Nonaka and von Krogh's (2009) 'knowledge based view of the firm' (KBV). KBV builds upon the resource-based view and assumes that, as all tangible assets lie outside the firm, competitive advantage is more likely to originate from internal firm-specific knowledge and the organization's ability to create new knowledge (Spender 1996).

Focusing on knowledge as a resource, Drucker (1959, 1993), Nonaka (1994, 2000, 2009), Spender (1996), von Krogh et al. (2001), Bratianu and Orzea (2010) and Donnelly (2019) each argue that it is the organization's ability to create, harness and utilise knowledge that significantly increases inimitable strategic assets. As well as exposing the relationship between knowledge and the generation of unique resources, the KBV has developed to encapsulate knowledge, knowledge assets, the context in which knowledge is created, and the way in which knowledge is utilised at individual, group and organizational level (Nonaka et al. 2000; Nonaka and Toyama 2003; Nonaka and von Krogh 2009). In doing so, KBV captures the links between knowledge as a resource and its strategic capacity to contribute to an organization's competitive advantage.

Table 1 summarizes the chronology and development of the KBV of the firm as a useful contextual tool to be used in formatting the intellectual boundaries within this chapter.

Table 1 reveals the importance of knowledge as critical in enhancing competitive advantage, the importance of tacit and explicit knowledge in the knowledge creation process, and the development of a model theorizing how tacit knowledge

Table 1 Knowledge Based View of the Organization

Component	Definition	Level
Knowledge as a resource	Organization's ability to create, harness and utilize knowledge and increase inimitable strategic assets	Organization
Knowledge as a strategic asset	Tacit and explicit knowledge differentiated and the organization identified as a place where knowledge is created, transferred and shared	Organization
Resource Based View	Competitive advantage lies in the application of rare and inimitable resources, competencies and capabilities which are scarce, valuable and durable	Organization
Knowledge Based View	Competitive advantage is created from the strategic use of intangible resources and 'firm specific knowledge' unique to the firm and unavailable to competitors	Organization
Knowledge conversion	Organizational knowledge is created through the continuous interaction between tacit and explicit knowledge	Individual

Table 2 Models of Knowledge

Knowledge	Overview
Intellectual Capital	Knowledge as an asset, resource or competency that is decomposed into objective elements (e.g. the knowledge based resource)
Category	Categorize knowledge into discrete elements (e.g. tacit, explicit; codified, uncodified)
Socially constructed	Knowledge is constructed through a mix of explicit programs and a process of social interchange

Adapted from: McAdam and McCreedy (1999)

converts to explicit form. The identification of tacit knowledge as a source of competitive advantage originates with Winter (1987) and is supported Kogut and Zander's (1992) research which distinguishes between tacit and codified knowledge while also suggesting that organizations are social communities where knowledge is created, transferred and shared among individual workers. This socially constructed model of knowledge management views knowledge as intrinsically linked with the social and learning processes within the organization (McAdam and McCreedy 1999), a view echoed in this book. Under this mantel, knowledge transfer is the process by which an organization leverages knowledge and information among members, thereby promoting learning and producing new knowledge or understanding (Doherty and Cormican 2017).

Taken collectively, these models of knowledge creation fall into three distinct categories, as summarized in Table 2.

This chapter considers elements of each model in terms of knowledge creation; it assumes that knowledge is socially constructed, resulting in an imitable knowledge based resource that, if successfully channeled, results in a sustainable advantage for the organization in which it resides. Notwithstanding the perspective that 'all models must be treated with caution' (McAdam and McCreedy 1999, p. 95); Table 2 reveals the importance of knowledge as critical in enhancing competitive advantage, the importance of tacit and explicit knowledge in the knowledge creation process, and the development of a model theorizing ways in which tacit knowledge converts to explicit form.

2.1 Knowledge Conversion—The Journey from Tacit to Explicit Form

This chapter adopts Polanyi (1966, p. 7) perspective that tacit and explicit 'aspects of knowing have a similar structure and neither is ever present without the other'. The concept of converting tacit to explicit knowledge assumes that tacit and explicit forms of knowledge exist (Levina 1999), and that tacit knowledge can be converted to explicit form (Nonaka 1994; Nonaka and Takeuchi 1995; Bolisani and Scarso

2000; von Krogh et al. 2000; Roberts 2000; Johannessen et al. 2001; Schulz and Jobe 2001; von Krogh et al. 2001). In order to comprehend each form of knowledge as key components in the knowledge codification process, these concepts are summarized below.

Tacit knowledge is widely discussed but rarely defined (Kogut and Zander 1992) and researchers often seek to explain the term through reference to Michael Polanyi's assertion (1966, p. 4) that tacit knowledge is present when 'we can know more than we can tell'. For example, although we might recognize a face among a million others, or recognize moods within a person's face, we cannot readily put it into words as to how we make these distinctions (Nonaka 1994). When contemplating the value of this knowledge in an organizational setting, Berman et al. (2002) provide the example of a baseball batsman who has 0.4 s reaction time to accurately hit a ball travelling almost 45 meters per second. They attribute the player's accuracy in hitting the ball to a tacit understanding of the pitchers' movements and suggest that the batsman is reacting to a pattern of movement, acquired through cumulative experience and practice. Leonard and Sensiper (1998) suggest that individuals with a wealth of experience find and identify problems more quickly than less experienced individuals through, firstly, an ability to examine a problem intuitively and possibly reframe the way the problem is being approached and, secondly, by applying a process of overlaying a mind pattern generated from previous experience on to an existing problem. Brown and Duguid (1991) demonstrate how the application of an individual's tacit knowledge, borne out of his or her previous experiences, is responsible for the ways in which actual workplace practices differ from the ways these workplace practices are represented through manuals, training programs, organizational charts and job descriptions.

Nelson and Winter (1982) refer to tacit knowledge as that which underlies a skilled performance, even when an individual is not fully aware of or able to articulate the detail of her/his performance at a later stage. Subsequent studies reinforce the complexity of tacit-explicit interplay in the workplace (Stenmark 2001; Kalling 2003; Dougherty 2004; Håkanson 2007; Romme et al. 2010; Ractham and Srisamran 2018; Oyemomi et al. 2019). These authors each contend that this ability to recognize patterns constitutes knowledge that is difficult to articulate and capture in an organizational setting.

Baumard (1999) concurs with those who argue that tacit knowledge constitutes knowledge that cannot easily be articulated and is not documented precisely because it possesses a quality of non-communicability. Cowan et al. (2000) suggest that tacit knowledge signifies personal knowledge which remains uncodified, while Hedlund et al. (2003, p. 118) describe it as 'common sense'. Roberts (2000) reiterates the inherent incommunicable nature of tacit knowledge suggesting that tacit knowledge is an individual's uncodified personal knowledge that cannot be classed as information or data, while others find that some, but not all tacit knowledge is articulable in a supportive environment (Tsoukas and Vladimirou 2001; Gourlay 2003; Gourlay and Nurse 2005; Gourlay 2006a, b; Håkanson 2007; Tell 2016; Ractham and Srisamran 2018).

While theorists struggle to define tacit knowledge, the term describes knowledge which is not easily expressed, is un-expressible or not transmittable (Ancori et al.

2000). Adopting a combination of these perspectives, tacit knowledge is defined within this chapter as;

> Personal, context-specific and difficult to formalize and communicate, deeply rooted in an individual's action, experience and skilled performance.

Baumard (1999, p. 22) describes explicit knowledge as something that is 'written, recorded, validated and protected by a firm; either by patent or secret'. Explicit knowledge is stored and retrieved quickly and efficiently communicated through print, electronic methods and other formal means (Smith 2001). While explicit knowledge is commonly codified in symbolic form, it is also represented in equipment or other physical artefact (Echajari and Thomas 2015). With knowledge in this explicit format, an organization shares and diffuses knowledge assets to solve similar or reoccurring problems and share knowledge across multiple departments and functions (Choo 1998). Put another way, 'explicit knowledge has a universal character, supporting the capacity to act across contexts' (Nonaka and von Krogh 2009, p. 636).

When considering how knowledge moves from tacit to explicit form, Baumard (1999) argues that an organization's competitive advantage relies on the implementation of tacit knowledge in practice but that, paradoxically, such knowledge is often unrecognized within that organization. Although an organization may have untapped wealth of tacit knowledge within its workforce, Kikoski and Kikoski (2004) suggest that many organizations remain unaware of the existence of tacit knowledge, may misunderstand its nature, and/or lack a system or process to facilitate the conversion of tacit to explicit knowledge. Therefore, the awareness and identification of tacit knowledge within the workforce, and the promotion of its articulation, capture and codification is crucial to the success of an organization's knowledge creation activities.

The knowledge taxonomy relating to tacit and explicit knowledge within an organizational setting is summarised in Table 3.

Although the ease of communication and storage constitute important advantages of explicit knowledge, the management of explicit knowledge brings challenges. Relying on an explicit knowledge base requires investment in, for example, information communication technology (ICT) infrastructure and requires a stable environment to allow knowledge sharing. These requirements are reinforced by modern organizations' challenging demands, volatile market conditions, and changing customer demands and competition which require constantly changing, dynamic methods of harnessing and storing explicit knowledge (Smith 2001). Thus, managing explicit knowledge can be costly and increasingly portable. The challenge of imitation is also omnipresent via various technological innovations and there is little research evidencing how explicit knowledge contributes to higher profits (Cowan et al. 2000). Therefore, due to its ease of communicability and replication, explicit knowledge does not always provide competitive advantage. In some cases, where rapid changes in market conditions, customer demands and/or advancing technologies have direct impact upon the transferability and usefulness of explicit knowledge, organizations find themselves overloaded with irrelevant, obsolete or impossible to distribute data and information.

Table 3 Knowledge taxonomy

Knowledge	Description	Conversion Benefits	Challenges
Tacit	Knowledge which is personal, context-specific and deeply rooted in an individual's action, experience and skills It is not be easily described or communicated	Conversion to explicit form in order to create competitive advantage Allows individuals to apply their own experiences to tasks	Organizations are unaware of tacit knowledge, do not understand it and do not possess a system that facilitates conversion Difficult to codify
Explicit	Codified knowledge which is easily transmitted in such as by verbal means or through diagrams, computer programs, print, electronic methods and information technologies	Is stored, retrieved efficiently, communicated and shared through print, electronic methods and other formal means Knowledge, in this format, is shared and diffused across multiple departments and functions	Requires investment in information communication technology and a stable environment to allow sharing Ease of transfer may result in knowledge being captured/imitated by competitors

This insight reinforces the value of codifying internal tacit knowledge to the organization, which, when thoughtfully implemented, may be difficult to imitate in other settings.

2.2 *Tacit and Explicit Knowledge as a Continuum*

While the benefits of tacit and explicit knowledge to an organization are clearly outlined above, the possibility of harnessing and separating tacit from explicit knowledge is widely debated by researchers. While some conceptualize tacit and explicit knowledge as different types of knowledge, others argue that explicit knowledge exists at one end of a knowledge continuum with tacit knowledge on the other end (Nonaka and von Krogh 2009). This perspective suggests knowledge has both explicit and tacit characteristics but more and less of each depending on where it falls on the continuum. Arguing how embodied knowledge enables initial expressions of knowledge, Nonaka and von Krogh (2009) suggest that, through articulation, word experimentation, use of concepts and the use of linguistic relationships, embodied tacit knowledge moves more towards the explicit end of the continuum. Others differentiate between that which can be articulated and that which cannot (Tsoukas and Vladimirou 2001; Gourlay 2003; Gourlay and Nurse 2005; Gourlay 2006a, b; Håkanson 2007; Tell 2016; Ractham and Srisamran 2018).

In this chapter, we adopt the position that some (but not all) tacit knowledge is converted to explicit form, once the correct enabling conditions are created. Although agreeing that all knowledge has elements of both tacit and explicit, we believe that much of the tacit knowledge present in individuals within the workplace can be

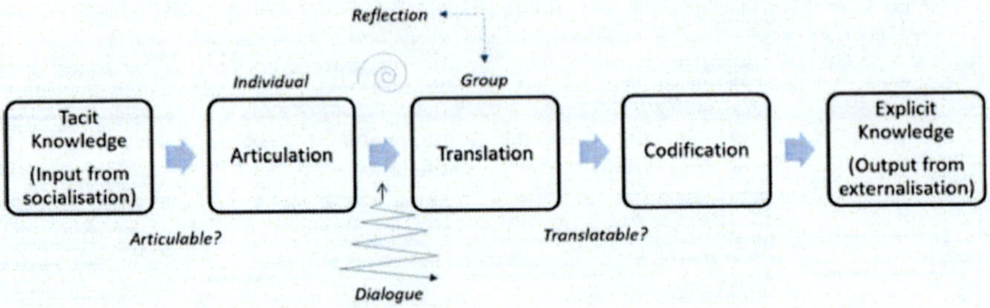

Fig. 1 Tacit to Explicit Knowledge Continuum

articulated and captured over time, once organizational tacit stores are identified and conditions facilitating externalization and codification exist (Fig. 1).

Once the concept of a continuum of knowledge is accepted (Fig. 1), the challenge of moving knowledge along the continuum from tacit to explicit form becomes pertinent. Noteworthy within the critique concerning the conversion of tacit to explicit knowledge is Polanyi's (1966, p. 5) proposition that knowledge is only transferable where an 'intelligent co-operation' between the receiver and the sender exists. This co-operation necessitates the receiver having prior knowledge, which he or she applies to the current message and context. Szulanski (1996) similarly argues that a challenge in sharing knowledge may be attributable to an individual's inability to apply outside sources of knowledge to a situation. Spender (1996) claims that some tacit knowledge will never be codified and cites cognitive and physical abilities as dimensions of tacit knowledge that can never be made explicit. Cook and Brown (1999) and Tsoukas (2003) argue that the conversion of tacit to explicit knowledge is impossible because knowledge is created through social practices and experiences which take place under the guidance of more experienced people. Therefore, the efficiency of knowledge conversion from tacit to explicit form depends on the sender and receiver possessing prior knowledge of the subject and context to which this knowledge relates. This suggests that the efficiency of tacit knowledge conversion improves where both sender and receiver share similar domain and contextual knowledge.

The argument as to whether tacit knowledge can be fully converted into explicit knowledge is developed by Håkanson (2007) who suggests that Polanyi's (1962, p. 51) examples of tacit knowledge such as bicycle riding and swimming are products of an individual's senses, perceptions and muscular control and cannot be converted to explicit form as simply as Nonaka suggests. Håkanson (2007) questions whether the tacit 'know how' disappears during the conversion of tacit knowledge to explicit knowledge. In a similar line of enquiry, (Soo et al. 2002) suggest that true knowledge is uncodified and that, as soon as knowledge becomes codified it ceases to be knowledge, instead becoming information and data, a view supported by recent research (Ractham and Srisamran 2018). We accept the limitations of codification in light of these insights, as discussed below.

2.3 Positions on Codification

Codification captures articulated tacit knowledge making it explicit and thereby transforming it into information that can be transferred, disseminated and archived (Ancori et al. 2000; Håkanson 2007; Kotlarsky et al. 2014; Kudryavtsev and Gavrilova 2017).

Two opposing positions regarding the codification of articulated knowledge exist. One view maintains that all knowledge is codifiable and that codification is simply the reduction of tacit knowledge into explicit form. The alternative view contends that some tacit knowledge is required for codified knowledge to be useful and that newly codified knowledge is a reconstruction of knowledge consisting of tacit and explicit elements containing a larger proportion of explicit knowledge than previously found (Ancori et al. 2000). This latter view of codification contends that, while codification aims to eliminate tacit knowledge, tacit knowledge remains necessary to decode and assign meaning to some structured data input (Balconi 2002). This latter position on codification aligns with Nonaka and von Krogh's (2009) theory contending that tacit and explicit knowledge exist at either end of a continuum and that tacit knowledge, through codification, is moved towards the explicit end of the continuum.

Taking the view of codification as a continuum, as knowledge emerges and is codified and re-codified, ambiguity and contextual facets lessen, reducing codified knowledge to a more systematic form (Cowan and Foray 1997). Articulated knowledge is codified at different levels of completeness ranging from codified knowledge that possesses high levels of tacit knowledge and is, therefore, only partly sharable, to highly explicated and sharable knowledge (Albino et al. 2001). Therefore, translating tacit knowledge into codified know-how changes the skills required to complete an activity (Balconi 2002). This is relevant in problem-solving where levels of tacit competence remain necessary to interpret explicated and decontextualized information and data in order to innovate (Balconi 2002).

Towards the explicit end of the tacit-explicit continuum, knowledge is codified without substantial change to its nature while, at the tacit end of the continuum, codified knowledge has no common basis with the original tacit knowledge (Malerba and Orsenigo 2000). While codified knowledge possessing low levels of contextual reference is easily shared, understood and used as intended by the transmitter (Bechky 2003b), codified knowledge that is completely decontextualized (Majchrzak et al. 2005) is difficult to apply appropriately at a later stage (Leonardi and Bailey 2008). Codification is complete once each tacit component is translated into a linguistic or graphical representation fully capturing the meaning of the original knowledge and its degree of 'completeness' is judged in terms of its representation of the original tacit meaning (Balconi et al. 2007, p. 833).

2.4 Defining Codification

Codification is variously defined as 'the process of conversion of knowledge into messages which can then be processed as information' (Cowan and Foray 1997, p. 596); 'the means by which knowledge is made explicit and hence readily stored or transferred between groups' (Kotlarsky et al. 2014, p. 609); 'the transformation of knowledge into information' (Ancori et al. 2000, p. 256); and 'the expression of knowledge in a standardized, fixed form' Håkanson (2007, p. 61). Codification captures articulated knowledge making it as explicit and portable as possible in order to maximize its accessibility (Kudryavtsev and Gavrilova 2017). Codification also captures and categorizes knowledge into explicit formats allowing the transfer, diffusion, storage and ready availability of knowledge (Håkanson 2007). In doing so, it facilitates the exposure of tacit knowledge within the organization, leading to knowledge and/or innovation breakthroughs (Fig. 2).

In Fig. 2, explicit knowledge is that knowledge already known to the organization that can be readily accessed through the codified knowledge channels in operation within the organization. This is 'the tip of the iceberg' relative to the potential tacit knowledge available to the organization. As exhibited in Fig. 2, the most valuable kind of knowledge for innovation and improvement and most likely to lead to breakthroughs resides beneath the surface. This knowledge is rooted in local settings, relying on the experience, practices and values of individuals within the organization. While difficult to communicate, tacit knowledge can be brought to the surface in the right communication environment as exhibited by the communication spiral, although its full value is not realized until added to the organization's codified knowledge (Fig. 2).

In this chapter, codification is the conversion of articulated tacit knowledge (Fig. 2) into information and messages which can then be transmitted and shared as procedures, guidelines, specifications or documents within the organization (Prencipe and Tell 2001; Whitaker et al. 2010).

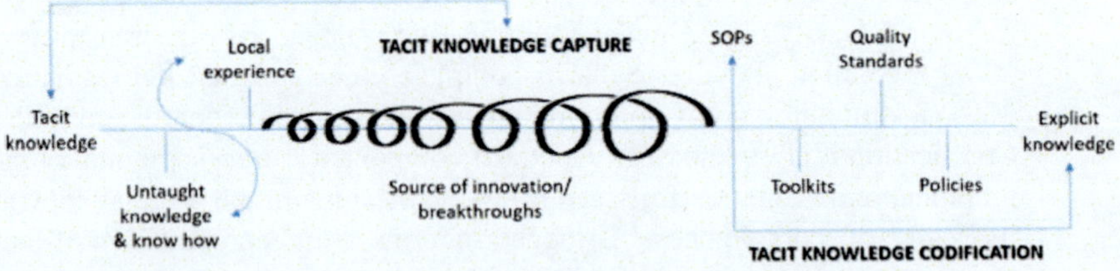

Fig. 2 Tacit to Explicit Knowledge enabled by the Codification Process

3 Capturing Articulated Tacit Knowledge

Tacit knowledge is articulated by personnel throughout the organization in a variety of environments wherein a range of techniques are employed. While factors such as respect, trust, recognition and open communication activities will free up people's minds to converse with each other, tacit knowledge conversion will be further enhanced in conversations through the deliberate and careful use of codified knowledge support material. Codifying undocumented articulated knowledge in this way allows us to capture it in a format that is beneficial to the organization and accessible and shareable by organizational stakeholders. In order to preserve, synthesize and develop this newly articulated knowledge, organizations should set out the expression of knowledge in a standardized fixed from which may exist as concepts, models, drawings, artefacts, or written language. Using codified material in this way allows conversationalists to optimize their dialogue and identify knowledge gaps and compare differences in opinion, understanding and perspectives in pursuit of newly articulated knowledge.

Once knowledge gaps are identified, they constitute opportunities when tacit knowledge can be articulated and/or developed. Therefore, using pre-existing codified materials both informs and directs conversation and functions as a means of furthering the articulation of tacit knowledge.

The value of informal codification techniques is that they can capture articulated knowledge in basic ways as a means of creating explicated materials with high tacit content. It is important to codify this knowledge in a variety of ways so as to make it accessible to as many people as possible. As a first step, a rudimentary codification document, drawing or artefact promotes dialogue which refines and expands the initial codification while also creating further articulation opportunities.

In addition to informal codification approaches, there are a variety of formal codification techniques that are of use when contemplating how to codify newly articulated knowledge in an organizational setting. These include; notes from meetings; suggestion logs or invited opinions; formal communication between shifts; mandatory use of email; analysis of observation or video recordings; creating and reviewing standard operating procedures; process mapping; creating troubleshooting guides; use of structured problem-solving methodologies; comparative exercises; cross-team analysis and the use of intermediaries to translate articulated knowledge into appropriately codified knowledge. Feedback forms are also used to embed open communications. These are in addition to regular brainstorming sessions and structured documentary tools such as standard operating procedures (SOPs) as exhibited in Fig. 2.

The various codification techniques are summarized in Table 4.

Involving operators in tasks such as writing documentation, designing layouts and value streams are key techniques that are used in the codification of knowledge. For example, process mapping or value stream mapping is a valuable means of capturing and codifying tacit knowledge residing within the organization. A key advantage of process mapping is that the organization can capture a lot of knowledge from the individuals directly participating in the process. In order to gain as much

knowledge as possible, the process mapping team should consist of people from diverse backgrounds and levels of knowledge, and should involve participants with in-depth knowledge of the process being deconstructed. Using a process mapping process as a means of accessing and codifying individual employee's knowledge also serves to identify small but perhaps important differences between standard operating procedures (SOPs) and what an operator may actually do. In this way, the value of process mapping as a means of generating as well as capturing knowledge can be captured. While outside of the realm of this chapter, the importance of process mapping as a means of involving and engaging a range of employees to articulate their knowledge should not be underestimated. Key to this process is recognition of the fact that individual knowledge and experience is articulated during the mapping process as a consequence of the interactive nature of the technique. The mapping process acts as a catalyst for employees to articulate and exchange knowledge. It is during the mapping process that people are encouraged to reflect on practice and challenges and propose ideas or highlight issues. As highlighted in the above example, this may result in the identification of deviations from SOPs which manifest the individual operators' tacit knowledge input to the process.

Alternatively, creating and in particular inviting operators to write and review Job Breakdown Sheet (JBSs) allows for a combination of describing, systematizing, justifying and questioning every element of the task or process under consideration and further facilitates the articulation, capture and codification of tacit knowledge. Transfer from articulation to codification is further enhanced if personnel are trained to write and review JBSs and process maps. Standard work methodologies, such as JBSs and analysis of video observation, enhance the articulation-codification cycle in a focused, structured and systematized way, allowing for tacit knowledge to be codified within the organization.

Codifying knowledge in expedient formats is a vital first step in ensuring that articulated tacit knowledge is successfully captured for use in the whole organization. While newly articulated knowledge is codified expediently initially and in a relatively basic manner, such as via a sketch, a few words hastily written onto a Post-it Note, a photograph or a few bullet points in an electronically created document, most of those experienced in this process then incorporate this initial knowledge capture into organizational documentation in more complex and formal ways. The most common means of incorporating newly articulated knowledge (codified in expedient but rather basic ways) is to update the existing SOPs to which the newly articulated knowledge pertains. A technique to support codification where an individual may be able to articulate but not to codify their knowledge is the provision of an intermediary who will translate articulated knowledge into appropriately codified knowledge.

Once articulated knowledge is codified into more sophisticated, accessible or standardized forms, the newly updated procedures are integrated into the organization's systems where they become shareable.

4 Embedding Codified Knowledge in the Organization

As discussed previously, codification conveys ideas, facts and processes and assists in creating shared understanding (Kudryavtsev and Gavrilova 2017). Due to increased tangibility and ease of access, codified knowledge positively affects rates of knowledge creation and innovation, directly affecting the generation and distribution of tacit knowledge (Cohendet and Steinmueller 2000; Kotlarsky et al. 2014), thereby reducing dependencies between diversified groups (Vaast and Levina 2006). Codification provides learning opportunities (Zollo and Winter 2002; Echajari and Thomas 2015) as is evident where project teams apply 'lessons learned' to new projects (Swan et al. 2010). Codification, undertaken as a competence building exercise, may also provide value additional to yielded codified knowledge in that benefits to the organization accrue, not only from the resulting codified knowledge, but also as a result of the critical and analytical activities that occur during the process (Zollo and Winter 2002).

Literature asserts that newly codified knowledge is most beneficial when embedded in boundary objects which bridge domains of specialist knowledge. These boundary objects provide a 'hook' upon which individuals with different perspectives 'hang' their contextual interpretations (Carlile 2002; Bechky 2003a, b; Carlile 2004). Pursuit of shared understanding and articulated knowledge using these boundary objects bridge contextual, social and cultural differences facilitating knowledge sharing between different contextual backgrounds (Carlile 2004; Akkerman and Bakker 2011).

Different formal codification methods can be used in this way including the creation of: suggestion schemes; video diaries or video files; standard job sheets/specifications (also known as job breakdown sheets); process mapping; troubleshooting guides, and the use of structured problem-solving methodologies, as described in Table 4. By pursuing an organic standardized work document infrastructure, new codified knowledge can be embedded in a uniform layout, contextual background and language. In doing so, recognizable, easily interpretable, accurate and comprehensive communication can be facilitated in ways that are common throughout the organization (Lee 2007). While standard work documents create shared understanding, they also provide de-contextualized knowledge with a contextual base making the knowledge more easily interpreted and understood by those familiar with the concept of standard work. In this regard standard work documents constitute tangible boundary objects. While acknowledging the different ways in which articulated knowledge is codified within different organizations (Table 4), most organizations use standard work documentation where each job is broken down to its basic elements so operation procedures are clear and concise. Thus, formal codification techniques that can capture articulated knowledge as standardized procedures create tangible boundary objects with high levels of explicated knowledge. This documentation enables organizations to, firstly, codify articulated knowledge in fixed forms and, secondly, give de-contextualized knowledge a contextual base which is universally understood throughout the organization.

Emerging from the discussion of informal and formal codification is the point that tangible boundary objects, created through informal and formal codification exercises, constitute boundary objects located at different points along the codification continuum. On such boundary object is the codebook.

4.1 The Concept of Codebooks

A codebook is a type of document used for gathering and storing codes. Originally codebooks were often literally books, but today codebook is a byword for the complete record of a series of codes, regardless of physical format. They address the challenges of ambiguity, inconsistency and shared understanding associated with codification (Cowan and Foray 1997; Cowan et al. 2000; Cohendet and Steinmueller 2000; Zollo and Winter 2002; Bechky 2003b; Bittner and Leimeister 2014). Thus, codification necessitates a codebook such that, once the correct codebook exists, all articulable knowledge is codifiable (Cowan et al. 2000).

Literature suggests that, in order to be useful, codebooks must be readable and understood by both sender and receiver, as codified knowledge with high tacit elements is only partially useful (Lissoni 2001; Albino et al. 2001). Thus, there is a need for codebooks to thoroughly capture the original meaning and intentions of those who have created them and, secondly, that codebooks must be intelligible to users. Furthermore, explicated tacit knowledge requires a commonality with its original tacit meaning otherwise it will not be shared or understood as intended by the original transmitter (Malerba and Orsenigo 2000). This suggests that levels of contextual reference must be removed from codified knowledge in order that individuals from different contextual backgrounds can access and apply it (Cowan et al. 2000; Albino et al. 2001; Bechky 2003b).

A codebook defines the way in which articulated knowledge is captured and understood and encapsulates commonly used and understood vocabulary, technical specification, managerial procedures, internal standardized work documents or drawings (Bénézech et al. 2001). In doing so, it provides 'a vocabulary of precisely defined and commonly understood terms, and a grammar to stabilize the language' (Cohendet and Meyer-Krahmer 2001, p. 1564). Codebooks also provide a grammar or language, in verbal or non-verbal form, that can be used consistently throughout the organization (Cohendet and Meyer-Krahmer 2001; Balconi et al. 2007; Eppler and Burkhard 2007; Kudryavtsev and Gavrilova 2017).

While they are often described as existing as text and language (Cowan and Foray 1997; Cowan et al. 2000; Steinmueller 2000), codified knowledge also manifests implicitly as international standards (such as ISO 9000) (Bénézech et al. 2001), external regulation and internal standardized work documents and job descriptions. Standard work documentation constitutes a codebook format which is intelligible and accessible to everyone required to use it and process mapping constitutes a different codebook format (O'Meara and Kelliher 2020). Process mapping constitutes structured social interactions where individuals, from diverse backgrounds, articulate and codify knowledge together. The practice and value of involving those working within the process aligns with theoretical recommendations emphasizing the importance of capturing contextual aspects between employees with different skill sets or experiences, bridging contextual differences and creating a common understanding (Star and Griesemer 1989; Carlile 2002, 2004; Akkerman and Bakker 2011).

Codified knowledge also exists within non-verbal forms (Balconi et al. 2007) such as concepts, models, drawings, procedures, artefacts images, sketches, conceptual diagrams, visual metaphors and interactive visualizations (Eppler and Burkhard 2007; Kudryavtsev and Gavrilova 2017; O'Meara and Kelliher 2020). These implicit and non-verbal manifestations of codified knowledge extend the term 'codebook' to describe codified knowledge within a variety of forms extending beyond text and language (Johnson et al. 2002).

One potential limitation is that knowledge codified in one context may not readily be understandable or easily applicable in another. Codebooks make codified knowledge accessible but render it useless to those without access to the code (Lissoni 2001). Therefore, codebooks must be understood by the creator *and* the 'reader' in order to be useful. To address the challenge of inconsistency and ambiguity within codification (Cowan and Foray 1997; Cowan et al. 2000) standardized codification practices and codebooks are required (Cohendet and Steinmueller 2000). If no established way of recording newly articulated knowledge exists either through language, drawings, standards or procedures, knowledge remains articulated but uncodified or only partially codified.

4.2 *The Value of Employee Engagement in the Codification and Recodification Process*

The effectiveness of standard work documentation and process mapping as codebook formats is attributed to the fact that both techniques enable articulated knowledge to be initially codified and then refined and re-codified through critical examination from multiple perspectives. This reinforces the importance of a capture system which allows articulated knowledge to be captured and then codified by a codebook which is understandable to all. Therefore, the review of standard work documentation by knowledgeable individuals involved in the process adapts existing codified knowledge within codebooks making them more 'complete' (Balconi et al. 2007) and creating newly codified knowledge that is located further towards the explicated end of the codification continuum.

Literature suggests that higher levels of knowledge sharing occur when employees perceive that their knowledge contributions are recognized and appreciated within the organization (Cabrera et al. 2006; Lin 2007). One such knowledge sharing method is through inviting operators to contribute knowledge through the creation and review of existing codified material. This invitation not only demonstrates recognition and appreciation of their knowledge contribution, it also enhances articulation by inviting

handbooks, software applications and other designated organizational information portals.

Promoting an open communications culture: Careful use of codified material to inform and direct conversation creates a reference point and shared understanding which leads to the identification of knowledge gaps and articulation.

Cross-division knowledge sharing: Applying the practice of rotating employees throughout an organization in order to open communication channels both within and across organizational silos is of particular value in embedding codified tacit knowledge across the organization. When carefully constructed cross-functional teams (i.e. teams consisting of interactive participants with a variety of knowledge specialisms, multiple perspectives, experiences, abilities and responsibilities) are tasked with formal codification exercises, using pre-codified material, articulated knowledge tends to be stimulated and can be immediately captured and codified.

Peer-review: Examining codified material for discrepancies against work practices, in conjunction with operators, encourages articulation and codification/re-codification. Newly articulated knowledge, relating to discrepancies is used to recodify existing material when the articulated knowledge is advantageous. When the articulated knowledge reveals disadvantageous practice, existing codified material (for example SOPs and standardized work documentation) can also re-codified but this time to emphasize specific points in order to prevent future discrepancies.

References

Akkerman, S. F., & Bakker, A. (2011). Boundary crossing and boundary objects. *Review of Educational Research, 81*(2), 132–169.

Albino, V., Garavelli, A. C., & Schiuma, G. (2001). A metric for measuring knowledge codification in organisation learning. *Technovation, 21,* 413–422.

Ancori, B., Bureth, A., & Cohendet, P. (2000). The economics of knowledge: The debate about codification and tacit knowledge. *Industrial and Corporate Change, 9*(2), 255–287.

Balconi, M. (2002). Tacitness, codification of technological knowledge and the organisation of industry. *Research Policy, 31,* 357–379.

Balconi, M., Pozzali, A., & Viale, R. (2007). The "Codification Debate" revisited: A conceptual framework to analyse the role of tacit knowledge in economics. *Industrial and Corporate Change, 16*(5), 823–849.

Baumard, P. (1999). *Tacit knowledge in organizations*. London: SAGE.

Bechky, B. A. (2003a). Object lessons: Workplace artifacts as representations of occupational jurisdiction. *American Journal of Sociology, 109*(3), 720–752.

Bechky, B. A. (2003b). Sharing meaning across occupational communities: The transformation of understanding on a production floor. *Organization Science, 14*(3), 312–330.

Bénézech, D., Lambert, G., Lanoux, B., Lerch, C., & Loos-Baroin, J. (2001). Completion of knowledge codification: An illustration through the ISO 9000 standards implementation process. *Research Policy, 30,* 1395–1407.

Berman, S. L., Down, J., & Hill, C. W. L. (2002). Tacit knowledge as a source of competitive advantage in the National Basketball Association. *The Academy of Management Journal, 45*(1), 13–31.

Bicheno, J. (2004). *The new lean toolbox: Towards fast, flexible flow*. Buckingham: Picsie Books.

Bittner, E. A. C., & Leimeister, J. M. (2014). Creating shared understanding in heterogeneous work groups: Why it matters and how to achieve it. *Journal of Management Information Systems, 31*(1), 111–143.

Bolisani, E., & Scarso, E. (2000). Electronic communication and knowledge transfer. *International Journal of Technology Management, 20*(1/2), 116–133.

Bratianu, C., & Orzea, I. (2010). Organizational knowledge creation. *Management & Marketing Challenges for Knowledge Society, 5*(3), 41–62.

Brown, J. S., & Duguid, P. (1991). Organizational learning and communities-of-practice: Toward a unified view of working, learning, and innovation. *Organization Science, 2*(1), 40–57.

Cabrera, A., Collins, W. C., & Salgado, J. F. (2006). Determinants of individual engagement in knowledge sharing. *The International Journal of Human Resource Management, 17*(2), 245–264.

Carlile, P. R. (2002). A pragmatic view of knowledge and boundaries: Boundary objects in new product development. *Organization Science, 13*(4), 442–455.

Carlile, P. R. (2004). Transferring, translating, and transforming: An integrative framework for managing knowledge across boundaries. *Organization Science, 15*(5), 555–568.

Choo, C. W. (1998). *The knowing organization: How organizations use information to construct meaning, create knowledge, and make decisions*. New York: Oxford University Press.

Cohendet, P., & Meyer-Krahmer, F. (2001). The theoretical and policy implications of knowledge codification. *Research Policy, 30*, 1563–1591.

Cohendet, P., & Steinmueller, W. E. (2000). The codification of knowledge: A conceptual and empirical exploration. *Industrial and Corporate Change, 9*(2), 195–209.

Cook, S. D. N., & Brown, J. S. (1999). Bridging epistemologies: The generative dance between organizational knowledge and organizational knowing. *Organization Science, 10*(4), 381–400.

Cowan, R., David, P. A., & Foray, D. (2000). The explicit economics of knowledge codification and tacitness. *Industrial and Corporate Change, 9*(2), 211–253.

Cowan, R., & Foray, D. (1997). The economics of codification and the diffusion of knowledge. *Industrial and Corporate Change, 6*(3), 595–622.

Dalkir, K. (2005). *Knowledge management in theory and practice*. Oxford: Elsevier.

Doherty, V., & Cormican, K. (2017). Analysis of knowledge transfer practices: Insights from a medical device manufacturing organization. *Procedia Computer Science, 121*, 186–193.

Donnelly, R. (2019). Aligning Knowledge Sharing Interventions With The Promotion Of Firm Success: The need for SHRM to balance tensions and challenges. *Journal of Business Research, 94*, 344–352.

Dougherty, D. (2004). Organizing practices in services: Capturing practice-based knowledge for innovation. *Strategic Organization, 2*(1), 35–64.

Drucker, P. F. (1959). *The landmarks of tomorrow*. New York: Harper & Row.

Drucker, P. F. (1993). *Post-Capitalist society*. Oxford: Butterworth Heinemann.

Echajari, L., & Thomas, C. (2015). Learning from complex and heterogeneous experiences: The role of knowledge codification. *Journal of Knowledge Management, 19*(5), 968–986.

Eisenhardt, K. M., & Santos, F. M. (2000). Knowledge-Based View: A New Theory of Strategy? In A. Pettigrew, H. Thomas, & R. Whittington (Eds.), *Handbook of strategy and management* (pp. 139–164). London: SAGE.

Eppler, M., & Burkhard, R. (2007). Visual representations in knowledge management: Framework and cases. *Journal of Knowledge Management, 11*(4), 112–122.

Gavrilova, T., Leshcheva, I., & Strakhovich, E. (2015). Gestalt principles of creating learning business ontologies for knowledge codification. *Knowledge Management Research & Practice, 13*, 418–428.

Gourlay, S. (2003) 'The SECI model of knowledge creation: Some Empirical Shortcomings. In *4th European Conference on knowledge management*. Oxford, England, 18–19 September, 2003 (pp. 1–10).

Gourlay, S. (2006a). Conceptualising knowledge creation: A critique of Nonaka's theory. *Journal of Management Studies, 43*(7), 1415–1436.

Gourlay, S. (2006b). Towards conceptual clarity for 'Tacit Knowledge': A review of empirical studies. *Knowledge Management Research & Practice, 4*(1), 60–69.

Gourlay, S., & Nurse, A. (2005). Flaws in the "Engine" of knowledge creation: A critique of Nonaka's Theory. *Challenges and issues in knowledge management* (pp. 293–315). Greenwich, Connecticut: Information Age Publishing.

Håkanson, L. (2007). Creating knowledge: The power and logic of articulation. *Industrial and Corporate Change, 16*(1), 51–88.

Hedlund, J., Forsythe, G. B., Horvath, J. A., Williams, W. M., Snook, S., & Sternberg, R. J. (2003). Identifying and assessing tacit knowledge: Understanding the practical intelligence of military leaders. *The Leadership Quarterly, 14,* 117–140.

Huber, G. P. (1991). Organizational learning: the contributing processes and the literatures. *Organization Science, 2*(1), 88–115.

Hutchings, K., & Michailova, S. (2004). Facilitating knowledge sharing in Russian and Chinese Subsidiaries: The role of personal networks and group membership. *Journal of Knowledge Management, 8*(2), 84–94.

Johannessen, J.-A., Olaisen, J., & Olsen, B. (2001). Mismanagement of tacit knowledge: The importance of tacit knowledge, the danger of information technology, and what to do about it. *International Journal of Information Management, 21*(1), 3–20.

Johnson, B., Lorenz, E., & Lundvall, B.-A. (2002). Why all this fuss about codified and tacit knowledge? *Industrial and Corporate Change, 11*(2), 245–262.

Kalling, T. (2003). Organization-internal transfer of knowledge and the role of motivation: A qualitative case study. *Knowledge and Process Management, 10*(2), 115–126.

Kikoski, C. K., & Kikoski, J. F. (2004). *The inquiring organization: Tacit knowledge, conversation, and knowledge creation: Skills for 21st-century organizations.* Westport: Praeger.

Kogut, B., & Zander, U. (1992). Knowledge of the firm, combinative capabilities, and the replication of technology. *Organization Science, 3*(3), 383–397.

Kotlarsky, J., Scarbrough, H., & Oshri, I. (2014). Coordinating expertise across knowledge boundaries in offshore-outsourcing projects: The role of codification. *MIS Quarterly, 38*(2), 607–627.

Kudryavtsev, D., & Gavrilova, T. (2017). From Anarchy to system: A novel classification of visual knowledge codification techniques. *Knowledge & Process Management, 24*(1), 3–13.

Lee, C. P. (2007). Boundary negotiating artifacts: Unbinding the routine of boundary objects and embracing chaos in collaborative work. *Computer Supported Cooperative Work (CSCW), 16*(3), 307–339.

Leonard, D., & Sensiper, S. (1998). The role of tacit knowledge in group innovation. *California Management Review, 40*(3), 112–132.

Leonardi, P. M., & Bailey, D. E. (2008). Transformational technologies and the creation of new work practices: Making implicit knowledge explicit in task-based offshoring. *MIS Quarterly, 32*(2), 411–436.

Levina, N. (1999). *Knowledge and organizations literature review.* Cambridge, Massachusetts: Massachusetts Institute of Technology.

Lin, H.-F. (2007). Effects of extrinsic and intrinsic motivation on employee knowledge sharing intentions. *Journal of Information Science, 33*(2), 135–149.

Lissoni, F. (2001). Knowledge codification and the geography of innovation: The case of brescia mechanical cluster. *Research Policy, 30,* 1479–1500.

Majchrzak, A., Malhotra, A., & John, R. (2005). Perceived individual collaboration know-how development through information technology-enabled contextualisation: Evidence from distributed teams. *Information Systems Research, 16*(1), 9–27.

Malerba, F., & Orsenigo, L. (2000). Knowledge, innovation activities and industrial evolution. *Industrial and Corporate Change, 9*(2), 289–314.

McAdam, R., & McCreedy, S. (1999). A critical review of knowledge management models. *The Learning Organization, 6*(3), 91–100.

Nelson, R., & Winter, S. (1982). *An evolutionary theory of economic change.* Cambridge Massachusetts: Harvard University Press.
Nonaka, I. (1991). The knowledge-creating company. *Harvard Business Review, 69*(6), 96–104.
Nonaka, I. (1994). A dynamic theory of organizational knowledge creation. *Organization Science, 5*(1), 14–37.
Nonaka, I., & Takeuchi, H. (1995). *The knowledge-creating company: How Japanese companies create the dynamics of innovation.* New York: Oxford University Press.
Nonaka, I., & Toyama, R. (2003). The knowledge-creating theory revisited: Knowledge creation as a synthesising process. *Knowledge Management Research & Practice, 1*(1), 2–10.
Nonaka, I., Toyama, R., & Konno, N. (2000). SECI, Ba and leadership: A unified model of dynamic knowledge creation. *Long Range Planning, 33*(1), 5–34.
Nonaka, I., & von Krogh, G. (2009). Tacit knowledge and knowledge conversion: Controversy and advancement in organizational knowledge creation theory. *Organization Science, 20*(3), 635–652.
O' Meara, M., & Kelliher, F. (2020). Exploring the process of codifying tacit knowledge in an organisational setting. In *Irish Academy of Management Conference Proceedings*, Aug 26
O'Meara, M., & Kelliher, F. (2017). Framing the externalisation phase of knowledge creation in an organisational setting. In *18th European Conference on Knowledge Management, Universitat Internacional de Catalunya*, Spain, Sept 7–8.
Oshri, I., Pan, S. L., & Newell, S. (2005). Trade-Offs between knowledge exploitation and exploration activities. *Knowledge Management Research & Practice, 3,* 10–23.
Oyemomi, O., Liu, S., Neaga, I., Chen, H., & Nakpodia, F. (2019). How cultural impact on knowledge sharing contributes to organizational performance: Using the fsQCA approach. *Journal of Business Research, 94,* 313–319.
Pathirage, C. P., Amaratunga, D. G., & Haigh, R. P. (2007). Tacit knowledge and organisational performance: Construction industry perspective. *Journal of Knowledge Management, 11*(1), 115–126.
Polanyi, M. (1962). *Personal knowledge: Towards a post-critical philosophy.* New York: Harper & Row.
Polanyi, M. (1966). *The tacit dimension.* New York: Doubleday.
Prencipe, A., & Tell, F. (2001). Inter-Project Learning: Processes and outcomes of knowledge codification in project-based firms. *Research Policy, 30*(9), 1373–1394.
Ractham, V. V., & Srisamran, P. (2018). Effects of knowledge articulation and self-reflection on team performance. *Knowledge Management & E-Learning, 10*(2), 177–195.
Roberts, J. (2000). From Know-How to Show-How? Questioning the role of information and communication technologies in knowledge transfer. *Technology Analysis & Strategic Management, 12*(4), 429–443.
Romme, A. G. L., Zollo, M., & Berends, P. (2010). Dynamic capabilities, deliberate learning and environmental dynamism: a simulation model. *Industrial and Corporate Change, 19*(4), 1271–1299.
Romme, G., & Dillen, R. (1997). Mapping the landscape of organizational learning. *European Management Journal, 15*(1), 68–78.
Schulz, M., & Jobe, L. A. (2001). Codification and tacitness as knowledge management strategies: An empirical exploration. *Journal of High Technology Management Research, 12*(1), 139–165.
Singh, J., & Singh, H. (2012). Continuous improvement approach: State-of-art review and future implications. *International Journal of Lean Six Sigma, 3*(2), 88–111.
Smith, E. A. (2001). The role of tacit and explicit knowledge in the workplace. *Journal of Knowledge Management, 5*(4), 311–321.
Soo, C., Devinney, T., Midgley, D., & Deering, A. (2002). Knowledge management: Philosophy, processes, and pitfalls. *California Management Review, 44*(4), 129–150.
Spender, J. C. (1996). Making knowledge the basis of a dynamic theory of the firm. *Strategic Management Journal, 17,* 45–62.

or assessing the usage of organizational knowledge, every phase in the KM process has been subject to widespread analysis and study.

As a business asset, knowledge has very special and paradoxical characteristics as its usage does not result in its consumption and with knowledge transfer also not leading to its loss. However, on the other hand, much of an organization's most valuable knowledge walks out of the door at the end of the day on every single day (Dalkir 2005).

In the new digital world, where simple systems and tools can allow anyone to quickly access so much data and information, how can we deploy the right information, in the right way and at the right time to, by converting simple data and information into the right knowledge and wisdom, increase our efficiency, our productivity and add value to our products and services, crucial to winning the struggle for competitiveness in the global marketplace?

Will managing the knowledge of our business or organization based on strong information technology (IT) systems, collecting and storing large amounts of data, as well as all of our documents and files in a very complex document management system, or even implementing a new talent recruiting and management strategy, be enough to foster and properly apply the organization's real knowledge?

We should be able to create and maintain so much important knowledge within our organizations. However, what is its purpose if individuals or their teams are not able enough to apply it to leading the respective market? Additionally, how can we manage and deploy all of the tacit knowledge of our highly mobile employees, or those on the cusp of retirement, which are commonly an essential part of the organization's competitive advantage if there is not the appropriate and timely leadership for sharing it?

When we are unable to share all this knowledge with our team and beyond, the purpose of such knowledge to organizations fails to extend beyond filling a large number of highly secure and sometimes very confidential folders, whether classified and organized in large and restricted-access rooms or stored in very expensive IT systems, with equally expensive IT experts managing them. In either case, only a few organization members will be able or allowed to access and use such knowledge for the creation of value to the business. In fact, knowledge will always be of only extremely low value whenever organizations lack the capacity to promote its sharing and application for the success of the business and its organizational goals.

However, is this an easy task? For many people, knowledge means power! And tacit knowledge (knowledge individuals do not know they have) is always going to be difficult to share! In these circumstances, how can organizations ensure their team members and partners share knowledge?

In accordance with the findings and principles set out above, our proposal in this paper involves discussing the importance of knowledge sharing to the competitiveness of any organization.

After this introduction to knowledge sharing, we describe some of the main barriers and incentives, followed by the identification of particular measures and strategies for overcoming the organizational barriers to improving the implementation of successful strategies through focusing on organizational culture and leadership styles.

2 Barriers and Incentives to Knowledge Sharing

2.1 *Sharing Knowledge Within the Knowledge Management Process*

Nowadays, the corporate memory is a real and important issue to any organizational strategies and practices. However, does this go far enough? With all the data, information and experiences stored in important and certainly expensive IT systems, this will not itself be enough to maintain healthy and profitable organizations faced with the competitiveness of global markets.

Any first approach to these problems certainly incorporates the importance of storage and the transfer of knowledge. However, organizations equally need to consider what results from all of this stored knowledge when nobody (or only a few) are able to actually access and use it. This especially applies when a part or even the entire workforce do not really understand the content/format.

In today's overly complicated world, success normally stems from the integration of different knowledge, skills and experiences. Indeed, this means not only deploying the data, experiences and knowledge of the past but also the ability to make it different, mostly (and adopting a buzzword of the moment) disruptively, extending to the continuous development and creation of new knowledge.

This only becomes possible due to all of the work and efforts of a very solid team in which all members use, discuss and especially share their own knowledge and experiences.

Organizations might be able to create, store and even transfer so much data, information and knowledge but, at the end of the day, whenever organizations are unable to apply and to transform all such knowledge and skills into added value for their clients, this is a pointless task and organizations end up losing money, opportunities, competitiveness and, as a final result, giving up market share.

Within the management cycle, knowledge transfers may be challengeable but this only constitutes an act of passage, from one person to another, a unidirectional act without any action or only brief participation by the recipient party. However, approaches to such sharing also perceive the transfer of knowledge in both directions (or in every direction), which means that in a sharing strategy both sides bring knowledge into the community. This hence reflects the reason sharing might also represent the right strategy for creating and acquiring new knowledge and the best approach to storing and reusing knowledge in very profitable ways across any organization.

There is a general consensus that knowledge sharing is critical to organizations seeking to deploy their knowledge as an asset to achieve competitive advantage (King

2006; Wang and Hu 2020; Ganguly et al. 2019). Bock and Kim (2002) consider knowledge sharing as the most important facet of knowledge management (KM), and Inkpen (2002) points out that "unless individual knowledge is shared throughout an organization, the knowledge will have a limited impact on organizational effect".

Knowledge interflows amongst individuals in working groups enable them to enhance their competencies and mutually generating new knowledge (Sveiby 2001; Wang and Hu 2020). This results in synergistic effects. Hence, social capital is created as those who share knowledge refine it through dialogue while those who receive knowledge learn from it. Furthermore, this also implies that organizations need to assist their employees in becoming better aware of their tacit knowledge.

However, what is knowledge sharing or how can we really define it? As happens with KM, there have also been difficulties in establishing a consensus around a single definition for knowledge sharing.

We may however define knowledge sharing as the actions by which employees diffuse relevant information to others across the organization (Bartol and Srivastava 2002), which reflects how the ultimate goal of sharing employee knowledge is its transfer to organizational assets and resources (Dawson 2001).

As such, we may conclude that the major focus of knowledge sharing falls on the individual who can explicate, encode, and communicate knowledge to other individuals, groups, and organizations while also extending to the teams that have become so prominent in management thought and practice (King 2006),

We must also point out the sharp distinction between knowledge sharing and knowledge transfer, mainly because transfer implies a focus, a clear objective, and unidirectionality, while knowledge may be shared in unintended ways, multiple directionally, without any specific objective (King 2006), such as when teams attempt to develop mutual knowledge, common ground or establish what knowledge the parties know they share in common (Cramton 2001; Ganguly et al. 2019).

Hence, knowledge sharing among individuals within teams is a particularly important focus, whether these teams are temporary sets of interdependent individuals bound by a collective aim, problem solving groups (also usually temporary in nature), self-managing teams, or cross-functional teams (Glassop 2002; Wang and Hu 2020). Virtual teams, those in which individuals primarily communicate through electronic means, are now also becoming a more important focus of knowledge sharing.

There is widespread agreement that knowledge assets are difficult to replicate and constitute fundamental sources of competitive advantage in open economies. Furthermore, company competitive advantages seem increasingly predicated on the ability to identify and share knowledge to ensure its effective exploitation (Teece et al. 1997; Wang and Hu 2020).

Research in the field of knowledge sharing and transfer (Szulanski 1996; Jensen and Szulanski 2004; Ganguly et al. 2019) indicates that sharing and transferring knowledge processes are both very difficult and sticky. Szulanski (1995) introduced the concept of stickiness to knowledge transfer in order to highlight the difficulties involved. Stickiness emerges as an important determinant of the degree of diffusion and application of superior knowledge and more broadly of the abilities of companies to grow and prosper by replicating existing assets and capabilities (Szulanski 1995).

2.2 Barriers to Knowledge Sharing

Should knowledge be such a crucial asset to any organization and when sharing is so essential to their competitiveness, why is it so difficult to implement?

Many researchers, organizations and managers that strive to encourage knowledge sharing have focused on how they might best motivate individuals to share their most valuable personally held knowledge. However, we must first grasp the reasons people and organizations do or do not share their knowledge.

Knowledge sharing barriers are caused by social factors, technology issues and a combination of the two. The barriers related to social issues divide into two major categories: barriers attributable to the individual and barriers attributable to the organization.

At the individual level, the power of knowledge, trust, motivation and job security are crucial issues for bringing about successful knowledge sharing.

Some people presumably tend to share knowledge just as some people tend to be talkative. Others, nevertheless, follow the *knowledge is power* dictum, probably learned in organizational settings; these people may hoard knowledge and be reluctant over sharing.

Wah (2000) states that a major obstacle to KM is the propensity of people to *hoard knowledge*. Hoarding knowledge does seem to be natural, particularly under conditions of economic competition in which "knowledge is power". For example, sales staff may face quota pressures and strong competition from each other. Partial transfers of knowledge may be a more common type of hoarding in which sharers disclose selected circumstances about a case rather than the entirety of its information (Goh 2002).

One of the most important factors for enhancing knowledge sharing encapsulates the level of trust prevailing among co-workers. Most people are unlikely to share their knowledge and experience without a feeling of *trust* in the person in front of them. Any negative prior experiences with knowledge sharing shall certainly impact on their willingness to share their knowledge. Last but not least, the intrinsic motivation of the employee is another factor. They need to trust people will not misuse their knowledge while also trusting that the information received is accurate and credible due to its source.

The level of trust existing between the organization, its subunits and its employees greatly influences the amount of knowledge that flows both between individuals and from individuals into the firm's databases, best practices achieves and other records (De Long and Fahey 2000; Ogunmokun et al. 2020).

Knowledge is power and may lead to inequalities in status. Sharing one's knowledge can bring about a perceived *lack of job security*. People may regard sharing their knowledge and experience as weakening their corporate position and power within the company. Working environments often contain the fear that sharing their knowledge reduces individual job security because of the uncertainties about the intent of those with whom they are sharing their knowledge. Organizations may also contain employees that intentionally take ownership of their knowledge and experience so they receive recognition from their colleagues and peers.

Also of interest are the concepts and issues put forward by Yang and Wu (2007) to discuss what they identify as the "the social dilemma and prisoner's dilemma". Accordingly, the motivations for individual behaviour deserve first consideration and with the sharing of knowledge a personal behaviour. Based on different motivations such as competition, reciprocity, reputation, ego satisfaction, organizational climate and so forth, people may or may not share knowledge with others. In economics, the primary personal motivation of behaviour is self-interest. People do their best to maximize individual utility. When the knowledge they own is valuable to themselves, they are unlikely to share with others.

According to Miloff and Vogelstein (2012), there are many natural barriers to people and organizations sharing knowledge, specifically:

- Failure to appreciate the value of sharing knowledge,
- Lack of understanding how to effectively share knowledge,
- Lack of incentives or rewards (material or psychic) for knowledge sharing,
- Staff are busy and even with the best of intentions fail to develop knowledge sharing habits,
- Professionals are afraid to reveal they do not know something; do not want to take risks or be shown wrong out of embarrassment,
- Concern that sharing knowledge will reduce one's own value, prestige or recognition. Competition—real or perceived—for limited resources decreases the motivation for and the safety in sharing,
- The perceived benefits of knowledge hoarding make people feel secure, safe or powerful; people hope to benefit (financially, in power and credibility) from having exclusive access to knowledge,
- Lack of clarity over issues of confidentiality may lead to either withholding otherwise helpful information or sharing it inappropriately.

Some of the key examples of knowledge barriers involving individuals as identified by different authors are:

- The view of knowledge, whether as experience or as expertise, as a source of individual power (Disterer 2001; Ardichvili et al. 2006; Castellani et al. 2019),
- A lack of time either for sharing knowledge with colleagues or for identifying colleagues in need of knowledge (Riege 2005; Castellani et al. 2019),
- Trust related concerns, such as the fear that colleagues may take credit for knowledge shared by an individual or that an item of knowledge may either not be reliable or from a credible source (Riege 2005), and
- Lack of any motivation to share knowledge, such as understanding the benefits this may bring (Disterer 2001; Castellani et al. 2019).

Organizations have also taken different approaches to knowledge sharing. Some, believing in dangers of disclosing secrets or in viewing sharing as a diversion from the primary tasks of individuals, have never encouraged sharing. Others have rendered support out of a belief in the great potential benefit in disseminating knowledge within an organization and perhaps beyond its boundaries. Of course, the tenets of KM presume that sharing is generally both beneficial and necessary to organizations attaining their potential.

The knowledge barriers attributable to the organization range from the physical layout of working areas to hierarchical organization structures, which thwart knowledge transfers across functions and between hierarchical levels (Disterer 2001; Riege 2005).

Some key examples of the knowledge barriers attributable to the organization are:

- The observation that explicit knowledge tends to be shared to a greater degree than tacit knowledge, inhibiting the spread of certain knowledge types, e.g. experience (Riege 2005),
- Culture and background (Riege 2005; Ardichvili et al. 2006), language differences (Disterer 2001; Riege 2005; Ardichvili et al. 2006), gender differences (Riege 2005), and levels of education (Riege 2005).

Barriers related to technology often involve the very technology intended to facilitate knowledge sharing. According to Riege (2005), such barriers include:

- poor integration of IT systems and processes which compromise workflows,
- incompatibilities between different information systems,
- tardy maintenance of communication and collaboration support systems,
- inadequate user training,
- IT systems that fail to meet user requirements,
- the reticence of people over using systems they are unfamiliar with,
- overzealous expectations of a particular technology,
- and, an absence of organizational efforts to "sell" the benefits of the system to its users.

In extremely competitive environments, the tendency for people and organizations to build barriers to sharing their own knowledge is understandable. Knowledge is power and it is natural human behaviour for people and organizations to tend to protect themselves against new entrances into their market and to counter, whether individually or collectively, their competitors by saving all their knowledge as secrets and confidentialities in "strong safety boxes" available only to a select few.

3 Overcoming Organizational Barriers to Knowledge Sharing

3.1 Incentives and Motivation for Knowledge Sharing

In the global market, yesterday's knowledge is never enough. New and innovative knowledge needs to undergo continual production and application to keep organizations sustainable. Furthermore, this only becomes possible in an appropriate environment in which knowledge sharing forms the basis of the organization's culture and leadership.

As already set out, knowledge sharing implies a predisposition to interact with third parties over accessing knowledge considered useful or necessary for a given situation. From the literature and research review, we learn that knowledge sharing fundamentally occurs through interactions among groups (Nahapiet and Ghoshal 1998; Castellani et al. 2019).

Riege (2005) maintains that the efficient and focused sharing of pertinent knowledge results in faster learning for both the organization and the individual.

A remarkably interesting study developed by Yang and Wu (2007) relates to appropriate leadership roles and the collaborative culture for knowledge sharing. According to Yang and Wu, "successfully knowledge sharing or knowledge transferring relies on neither document nor information techniques. Knowledge sharing in an organization involves interactions between people. Under these conditions, an individual would consider the trade off between individual and organizational interests when making the decision to share knowledge with others. Since knowledge is powerful and a scarce resource in a knowledge-intense firm, people possessing important knowledge about an organization occupy a strong position and acquire some benefits in an organization. If people share their knowledge with others, their current advantages might suffer, or even be transferred to others. A rational individual would not easily share knowledge with others in these circumstances" (Yang and Wu 2007: 4).

Furthermore, individual behaviours also affect the utility of opponents and the final results of any team or organization. Hence, research findings conclude that organizational environments must provide a stimulus for employees to become active agents and contribute to the development of organizational knowledge. Assuming people are driven by motivations of self-interest, they "are likely to share their knowledge with others in an organization if they can gain additional payoff through doing so. People potentially benefit when they share knowledge with each other, by gaining or creating new specific knowledge, thus increasing the exclusive benefit in an organization. Due to the possibility of obtaining additional payoff, people are driven to share their valuable specific knowledge, even if there is a potential risk of losing their own knowledge advantages" (Yang and Wu 2007: 5).

In the contemporary world, any profitable organization needs to consolidate its competitiveness by instigating innovation processes in which people share their knowledge unselfishly so that this gets utilized effectively.

This requires a appropriate *organizational culture* and a strong *leadership style* in knowledge sharing processes, supported both by communication and trust and motivation and collaboration, which are primary and critical components to "unlocking" and stimulating sharing processes.

Communities of Practices (CoP) represent one example of how knowledge sharing may enhance and consolidate organizational knowledge as well as develop new and innovative knowledge.

3.2 A Supportive Organizational Culture

Placing the focus on fostering collaborative climates, in which collaboration encapsulates 'mutually sharing norms of behaviour', Yang (2007) starts out in his study by recalling that organizational culture was described by Robbins and Barnwell (1994) as the shared values, beliefs or perceptions held by employees within an organization or organizational unit that are not only agreed on by a significant proportion of members but also largely taken for granted by them. As expressed by the author, "culture is socially learned and transmitted by members, and can be found in any fairly stable social unit, of any size, as long as it has a reasonable history. In summary, culture provides norms/rules for behaviour in organizations."

An effective organizational culture is a key component influencing the capacity of organizations to survive and succeed over the long term (Schneider et al. 1994). Gupta et al. (2000), and Al-Kurdi et al. (2020) maintain that an organizational culture containing themes of openness and incentives successfully facilitates the integration of individual competencies (including skills, knowledge and experiences) into organizational knowledge through learning and knowledge creating and sharing.

According to many different studies, fostering the following components of organizational culture is susceptible to accelerating the implementation of KM practices:

- a 'collaborative, not a competitive' climate (Cameron 2006; Goh 2002; Ruggles 1998; Sveiby and Simons 2002; Oyemomi et al. 2019; Al-Kurdi et al. 2020),
- a trusting and trustworthy work environment (Goh 2002; Rowley 2002; Soliman and Spooner 2000; Sveiby and Simons 2002; Wenger et al. 2002),
- top management commitment (Hislop 2010; Mrinalini and Nath 2000; Rowley 2002; Oyemomi et al. 2019),
- mentoring programs (von Krogh 1998),
- accountability for sharing within a team (Bollinger and Smith 2001; Sawhney and Prandelli 2000),
- a focus on innovation, problem-seeking and problem-solving (Goh 2002; Oyemomi et al. 2019), and
- an opportunity for spontaneous and voluntary sharing (Dixon 2002).

Sveiby and Simons (2002) describe how the development of information systems and technologies fails to be successful without an individual willingness to share.

These authors identify the two major impediments to sharing as an "internal culture of resistance to sharing" and "a culture of hoarding knowledge". The components of 'collaboration' and 'trust' must be incorporated into the organizational culture for successful KM practices, focusing this culture on "the values, beliefs and assumptions that influence the behaviours and the willingness to share knowledge."

They identify the following 'enabling' characteristics of organizational culture:

- fostering trust in the workplace,
- encouraging knowledge sharing in action not words,
- promoting the introduction of new knowledge into the organization and developing insights and innovations for future success,
- stimulating employees to say what they think, and
- building open communication channels throughout the organization.

However, implementing knowledge sharing practices requires consideration of certain aspects of organizational culture (McDermott and O'Dell 2001), in particular:

- highlighting the relationship between KM initiatives and daily organizational routines. This means making visible the link between knowledge sharing and the practical results in terms of achieving business objectives and/or solving concrete problems,
- making knowledge sharing a "natural process" requires understanding just how information and knowledge exchanges really occur. Hence, visible KM activities – events, meetings, language, web sites – need to reflect organizational habits and routines even when the intention may be to promote new behaviours or approaches,
- defining values for knowledge sharing consistent with organizational values and not expecting people to share their ideas and contributions simply out of moral obligation. Once again, the language and practices adopted must be consistent with these values,
- fostering a culture of sharing through existing networks of relationships and interactions. This involves providing the tools and resources necessary for sharing to occur during these interactions and, above all, legitimizing these relational networks as a critical factor of success for knowledge sharing,
- seeking the support of people in the organization who stand out for sharing ideas and knowledge or influential managers who can influence the behaviours of other organization members. These people may serve as a type of "knowledge activist",
- KM initiatives also need to align and be consistent with other organizational practices (e.g. human resource management practices).

By its nature, organizational culture should be supportive and an enabler of knowledge sharing behaviour within an organization. Especially when the top management set the example, this becomes an enabler of motivations to share knowledge with other people within the workplace (Wang and Noe 2010; Al-Kurdi et al. 2020).

3.3 Leadership Styles

However, organizational culture is not in itself enough. Organizational culture and leadership are always in mutual accompaniment. Any cultural environment seeking progress will not survive without a strong leader coupled with total senior management commitment to teamwork and engaged in improving knowledge sharing just as no leader can lead successfully without the support of an organizational culture for teamwork and knowledge sharing.

Team leaders play important roles in nurturing healthy working atmosphere among their subordinates (e.g. Grandori and Kogut 2002; Hendriks 1999; McDermott and O'Dell 2001; Castellani et al. 2019; Moreno et al. 2020). The traditional view of management states that organizational members act as the instruments of their superiors (Roth 2003). However, this perspective is no longer seen as appropriate to securing long-term success and managers are increasingly required to stimulate subordinates to voluntarily transfer talent and experience into organizational assets. This involves leadership rather than management and facilitating and coaching roles correspondingly requiring more attention (Roth 2003). The studies of Chourides et al. (2003), Goh (2002) and Castellani et al. (2019) demonstrate how coaching leadership roles can positively facilitate KM. Mentoring programs enable senior members to assist juniors with senior staff members needing motivation to share their knowledge and experience with juniors and newcomers (von Krogh 1998).

Several scientific studies have concluded that leadership provides a crucial factor for facilitating processes of knowledge creation and sharing (von Krogh et al. 2012). In understanding leadership as the ability to influence the organization's members over achieving for the KM strategic objectives, it becomes the responsibility of organizational leaderships to promote the dissemination of their vision for KM through theirs commitment to knowledge sharing objectives and initiatives (McDermott and O'Dell 2001; Moreno et al. 2020). According to Kukko (2013), leadership plays a fundamental role in knowledge sharing, especially when firms are able to integrate the KM purposes into organizational goals. Senior managers must be the creators of a knowledge sharing culture within their strategic leadership as well as responsible for supporting knowledge development initiatives by facilitating the resources and means necessary for the implementation of these initiatives (von Krogh et al. 2012). Hence, leadership activities must also involve structural components that enable the management and development of the organization's knowledge assets (Nonaka et al. 2000a, b; Moreno et al. 2020), including aspects such as facilitating communication through information technologies (ITs), organizational structures and routines (promoting efficient horizontal and vertical communication channels) and providing the physical spaces for meetings.

However, the role of leadership in facilitating knowledge sharing is not limited to senior management. Nonaka (1994) maintains that a "top-down" perspective is not the most appropriate for KM. The top management does play a central role in promoting the KM vision but intermediate managers (supervisors, department and unit managers, project managers) are essential to stimulate shared contexts and boost the knowledge development processes (Nonaka et al. 2000a, b; Moreno et al. 2020) especially because they are the effective link between vertical and horizontal information and communication flows.

Approaches to leadership normally consider this a position, process or activity controlled by a central authority (von Krogh et al. 2012) as easily understandable from different studies on strategic leadership (and leadership styles) undertaken by various researchers (e.g. Lakshman 2007; Srivastava et al. 2006). However, a distinctive way of perceiving leadership involves assuming the distribution (and sharing) of control, authority and influence among different team members. Within this framework, leadership emerges through the mutual and constant interactions between the behaviours both of leaders and of all other team members (Drath et al. 2008), through dialogue and attribution, in what Drath and colleagues termed as "a view of leadership as dialogue and sense-making (Drath et al. 2008: 651)". In keeping with the dynamic nature of KM processes and their dependence on contextual factors (cultural and structural), knowledge sharing is likely to be best facilitated by leaders able to reconcile centralized and shared leadership styles (Nonaka et al. 2000a, b; von Krogh et al. 2012; Moreno et al. 2020).

Facilitating the leadership of learning improves an organization's ability to absorb knowledge (Simonin and Özsomer 2009), Facilitator leaders, assuming their roles as coaches or mentors, focus on the development of those around them, encouraging them to overcome barriers to learning, delegating responsibilities and motivating employees. Yang (2007) reports similar results in terms of the positive correlation between the "facilitator" and "mentor" leadership styles and the effectiveness of knowledge sharing. In this study, this positive correlation also extends to the "innovative" leadership style. In contrast, leadership styles involving excessively rigid policies and procedures negatively affect knowledge sharing. According to this same author, control and overly strict rules negatively influence the sharing of knowledge and with employees potentially perceiving them as coercive factors with associated punishments, understood and anticipated as the individual costs associated with knowledge sharing.

Furthermore, learning processes in an organizational context often occur based on "trial-error". Commonly, positive attitudes towards error associate with positive effects in terms of collaborative work and reinforcing a culture of sharing (e.g. Lindner and Wald 2011). Thus, managers should foster their openness to new ideas and changes (Cabrera et al. 2006; Sun and Scott 2005), perceiving at errors as development and learning opportunities rather than punishing team members for such mistakes.

Despite the importance of shared leadership in favouring formal relationships associated with experimentation and the sharing of personal experiences, this does not exclude the need for formalized roles or leadership activities. Both strands complement and reinforce each other and structural components such as vision, job layout and the formal definition of roles and responsibilities are able to reinforce the prevailing cultural values and collaborative working practices.

3.4 Communities of Practice

Communities of Practice (CoPs) have been highlighted as an effective method for knowledge sharing in KM practices and strategically applied by many organizations (Kim et al. 2012) as powerful mechanism for improving knowledge sharing among project managers, both within and between organizations (Lee et al 2015). CoPs receive particular recognition for their promotion of knowledge sharing within the scope of improving organizational innovation (Harvey et al. 2013).

The "*communities of professionals*" and "*communities of practice*" (CoP's) concepts provide a splendid conceptual instrument for recognizing how groups of professionals organize themselves at work, how they share knowledge, and how they learn, innovate and collaborate in ways that organizations often ignore in their descriptions of functions, professional training curricula and evaluation rules.

Organizations may draw significant value from them whenever actually prepared to recognize the existence of such communities, analyze their functioning, and create mechanisms, procedures and online tools that facilitate and improve the work of each such community. This should also target the innovation potentially resulting from their cross-sharing of explicit and tacit knowledge.

The issue of "professional communities", in the context of business activities, displays a long tradition, especially in the studies of industrial anthropology (Burawoy 1979) and the ethnography of organizations (Van Maanen and Barley 1984).

According to Van Maanen and Barley (1984) the "definition of an occupational community contains four elements, a group of people who consider themselves to be engaged in the same sort of work, who identify (more or less positively) with their work, who share with one another a set of values, norms, and perspectives that apply to, but extend beyond, work related matters, and whose social relationships meld the realms of work and leisure." However, not all occupations come to hold clearly decipherable contours as the degree to which knowledge, practices and values are shared among practitioners varies across occupations, across time, and across settings. "However, some occupations display a rather remarkable stability in social space and time and, hence, can be decoded. It is for them that the idea of an occupational community is most relevant since it draws attention to those occupations that transmit a shared culture from generation to generation of participants." (Van Maanen and Barley 1984).

Nevertheless, among many different authors, Wenger most contributed to the elaboration of the "CoP's" concept, evolving it from an analytical perspective to a prescriptive understanding and disseminating this throughout organizational and educational contexts (Wenger et al. 2002; Wenger 1998).

As Depalma (2009) describes: "Wenger provides this rigorous treatment, defining CoP's in terms of mutual engagement in a shared practice, negotiation of a joint enterprise, and the development of a shared repertoire"

And despite some criticism of Wenger's work, Depalma recalls how "Alinsu helped to make a useful point at the time: CoP's are indeed everywhere, even under the most dull and repressive regimes, whether managers want them or not. Yet this useful metaphor, as metaphors always do, involved a trade-off, and some of the breadth and complexity of Wenger's ideas have become lost in theoretical abstraction, overshadowed by the particular "reality" of Alinsu".

4 Conclusion

Knowledge is a key asset for any organization and knowledge sharing represents the best strategy both for creating more knowledge and for applying that existing to improve efficiency and add value to the company. However, with so many barriers to sharing knowledge, is this an easy task? No, it certainly is not!

Economic, behavioural and social factors all require consideration when assessing the issues around how to motivate individuals, or groups of individuals, to contribute their most valuable personal asset, transferring and sharing their own powerful knowledge to others who they may not even know as happens when contributing to KM systems.

Most of the interest and research on knowledge sharing have focused on this supply-side issue: that is, how to motivate people to share. However, some researchers have concentrated on the demand side: the knowledge seeking and knowledge-acquisition behaviours of individuals. This perspective addresses the potential users of knowledge and how they search for this when facing questions or problems. Expert networks have been established in organizations to enable such searches and with communities of practice clearly also facilitating this demand-side viewpoint of sharing.

No matter what individuals are apt to misunderstand, forget, filter, ignore and/or fail to pass on or whether this kind of withholding behaviour is unintentional or deliberate, organizational performance may nevertheless be affected. The incomplete transfer of knowledge incurs 'knowledge depreciation' or organizational forgetting even while deterring hoarding behaviours seems difficult. Inspiring individuals to share thus becomes crucial and organizations have to maintain healthy collaboration based climates.

Furthermore, the issues around how best to motivate individuals to share their most valuable personal knowledge are not completely resolved. Conventional wisdom states that nurturing a knowledge-sharing culture provides the best means even though this is not empirically well validated.

Among the other research findings on knowledge sharing that appear to achieve some consensus are the following:

- Knowledge sharing involves both costs and benefits (not necessarily economic),
- Contrary to certain popular wisdom, supervisory control appears to be more important than perceived organizational support in terms of both the frequency of submissions and the perceived effort expended on contributing to KM systems,

- Concerns over self-interest generate a negative effect on sharing-related attitudes. This might suggest that organizations fostering highly competitive cultures, such as policies of counselling 10% of the lowest performers out of the organization each year, might have difficulties in motivating knowledge sharing,
- Dispersed (not concentrated) computer-mediated teams encounter difficulties in knowledge sharing that are greater than those experienced in concentrated teams in part because of the difficulties in establishing a social presence — the degree to which the medium facilitates awareness of other people and the development of interpersonal relationships,
- Systems variables, such as use and usefulness, appear to have important moderating effects on individual sharing behaviours through KM systems.

While is no doubt about the importance of knowledge to the competitiveness of any organization, that knowledge only attains importance when able to apply it to the continuous improvement of process efficiency and in adding value to the marketed products and services.

Furthermore, in the contemporary world, yesterday's knowledge is never enough in the global market context and we therefore also need to be able to continuously create new and innovative knowledge.

However, knowledge is power and its sharing becomes so difficult as a direct result alongside the many barriers and walls that require overcoming.

As knowledge is power, just how might it be effectively shared? Trust, motivation, collaboration are the core drivers for a supportive organizational culture coupled with effective leadership successfully nurturing the sharing of knowledge within the organization.

Communities of practice as groups of people who share concerns or passion in something they do, regularly interacting to learn how to do this better, are also powerful vehicles for promoting a culture of learning and sharing of knowledge within domain of expertise and capable of providing benefits whether for organizations or each member/participant.

"***Build bridges. Not walls!***", Pope Francis said. Let us indeed build bridges instead of walls to share our knowledge and improve the quality of our own lives as well as that of our societies.

References

Al-Kurdi, O. F., El-Haddadeh, R., & Eldabi, T. (2020). The role of organisational climate in managing knowledge sharing among academics in higher education. *International Journal of Information Management, 50,* 217–227.

Ardichvili, A., Maurer, M., Li, W., Wentling, T., & Stuedemann, R. (2006). Cultural influences on knowledge sharing through online communities of practice. *Journal of Knowledge Management, 10*(1), 94–107.

Bartol, K., & Srivastava, A. (2002). Encouraging knowledge sharing: the role of organizational reward systems. *Journal of Leadership & Organizational Studies (JLOS).*

Bock, G., & Kim, Y. (2002, Apr–Jun). Breaking the myths of rewards: an exploratory study of attitudes about knowledge sharing. *Information Resources Management Journal (IRMJ), 15*(2), 14–21.

Bollinger, A. S., & Smith, R.D. (2001). Managing organizational knowledge as a strategic asset. *Journal of Knowledge Management, 5*(1), 8–18.

Cabrera, Á., Collins, W. C., & Salgado, J. F. (2006). Determinants of individual engagement in knowledge sharing. *The International Journal of Human Resource Management, 17*(2), 245–264.

Cameron, K. S., Quinn, R. E., Degraff, J., & Thakor, A. V. (2006). *Competing values leadership.* Cheltenham, England: Elgar.

Castellani, P., Rossato, C., Giaretta, E., & Davide, R. (2019). Tacit knowledge sharing in knowledge-intensive firms: the perceptions of team members and team leaders. *Review of Managerial Science,* 1–31.

Chourides, P., Longbottom, D., & Murphy, W. (2003). Excellence in knowledge management: an empirical study to identify critical factors and performance measures. *Measuring Business Excellence, 7*(2), 29–45.

Cramton, C. (2001). The mutual knowledge problem and Its consequences for dispersed collaboration. *Organization Science, 12*(3).

Cummings, J. N. (2004). Work groups, structural diversity, and knowledge sharing in a global organization. *Management Science, 50*(3), 352–364.

Dalkir, K. (2005). *Knowledge management in theory and practice.* New York: The MIT Press.

Dawson, R. (2001). Knowledge capabilities as the focus of organizational development and strategy. *Journal of Knowledge Management, 4*(4), 320–327.

De Long, D. W., & Fahey, L. (2000). Diagnosing cultural barriers to knowledge management. *Academy of Management Executive, 14*(4), 113–127.

Depalma, R. (2009). Leaving Alinsu: towards a transformative community of practice. *Mind Culture, 4,* 353–370.

Disterer, G. (2001). Individual and social barriers to knowledge transfer. In *Conference Proceedings 34th Annual Hawaii International Conference on System Sciences.* Los Alamitos, CA: IEEE Press.

Drath, W. H., McCauley, C. D., Palus, C. J., Van Velsor, E., O'Connor, P. M. G., & McGuire, J. B. (2008). Direction, alignment, commitment: toward a more integrative ontology of leadership. *The Leadership Quarterly, 19*(6), 635–653.

Ganguly, A., Talukdar, A., & Chatterjee, D. (2019). Evaluating the role of social capital, tacit knowledge sharing, knowledge quality and reciprocity in determining innovation capability of an organization. *Journal of Knowledge Management.*

Glassop, L. (2016). The organizational benefits of teams. *Human Relations, 55*(2), 225–249.

Goh, S. G. (2002). Managing effective knowledge transfer: an integrative framework and some practice implications. *Journal of Knowledge Management, 6*(1), 22–30.

Grandori, A., & Kogut, B. (2002). Dialogue on organization and knowledge. *Organization Science, 13*(3), 224–231.

Gupta, A. K., & Govindarajan, V. (2000). Knowledge management's social dimension: lessons from Nucor Steel. *Sloan Management Review, 42*(1), 71–80.

Gupta, B., Iyer, L. S., & Aronson, J. E. (2000). A study of knowledge management practices using grounded theory approach. *Journal of Scientific and Industrial Research, 59*(4), 668–672.

Harvey, C., Simon, & Dubois, (2013). Another cog in the machine: designing communities of practice in professional bureaucracies. *European Management Journal, 31*(1), 27–40.

Hendriks, P. (1999). Why share knowledge? The influence of ICT on the motivation for knowledge sharing. *Knowledge and Process Management. The Journal of Corporate Transformation, 6*(2).

Hislop, D. (2010). Knowledge management as an ephemeral management fashion?. *Journal of Knowledge Management, 14*(6), 779–790.

Inkpen, A. C. (2002). Learning through joint ventures: a framework of knowledge acquisition. *Journal of Management Studies, 37* (7), 1019–1044.

Jensen, R., Szulanski, G. (2004). Stickiness and the adaptation of cross-border knowledge transfers. *Journal of International Business Studies, 35*(6), 508–523.

Kim, S., Hong, J., & Suh, E. (2012) A diagnosis framework for identifying the current knowledge sharing status in a community of practice. *Expert Systems with Applications, 39*(18), 13093–13107.

King, W. (2006). Knowledge sharing. In D. Schwartz (Ed.) *Encyclopedia of knowledge management* (pp. 493–498).

Kukko, M. (2013). Knowledge sharing barriers in organic growth: a case study from a software company. *The Journal of High Technology Management Research, 24*(1), 18–29.

Lee, L., Reinicke, B., Sarkar, R., & Anderson, R. (2015). Learning through interactions: improving project management through communities of practice. *Project Management Journal, 46*(1), 40–52.

Lindner, F., & Wald, A. (2011). Success factors of knowledge management in temporary organizations. *International Journal of Project Management, 29*(7), 877–888.

Miloff, M., & Vogelstein, I. (2012). Barriers and benefits to knowledge sharing. Knowledge Communities. https://knowledgecommunities.com/page/2/.

Moreno, V., Cavazotte, F., & Dutra, J. P. (2020). Psychosocial and organizational antecedents of knowledge sharing in the workplace. *Revista de Administração Contemporânea, 24*(4), 283–299A.

Mrinalini, N., & Nath, P. (2000). Organizational practices for generating human resources in non-corporate research and technology organizations. *Journal of Intellectual Capital, 1*(2), 177–186.

Nahapiet, J., & Ghoshal, S. (1998). Social capital, intellectual capital, and the organizational advantage. *Academy of Management Review, 23*(2), 242–266.

Nonaka, I., Toyama, R., & Konno, N. (2000a). SECI, Ba and leadership: a unified model of dynamic knowledge creation. *Long Range Planning, 33*(1), 5–34.

Nonaka, I., Toyama, R., & Nagata, A. (2000b). A firm as a knowledge-creating entity: a new perspective on the theory of the firm. *Industrial and Corporate Change, 9*(1), 1–20.

Ogunmokun, O. A., Eluwole, K. K., Avci, T., Lasisi, T. T., & Ikhide, J. E. (2020). Propensity to trust and knowledge sharing behavior: an evaluation of importance-performance analysis among Nigerian restaurant employees. *Tourism Management Perspectives, 33,* 100590.

Oyemomi, O., Liu, S., Neaga, I., Chen, H., & Nakpodia, F. (2019). How cultural impact on knowledge sharing contributes to organizational performance: using the fsQCA approach. *Journal of Business Research, 94,* 313–319.

Riege, A. (2005). Three-dozen knowledge sharing barriers managers must consider. *Journal of Knowledge Management, 9*(3), 18–35.

Robbins, S., & Barnwell, N. (1998). *Organisation theory: concepts and cases.* Sydney: Prentice-Hall.

Rowley, J. (2002). Reflections on customer knowledge management in e-business. *Qualitative Market Research, 5*(4).

Ruggles, R. (1998). The state of the notion: knowledge management in practice. *California Management Review, 40*(3), 80–89.

Sawhney, M., & Prandelli, E. (2000). Communities of creation: managing distributed innovation in turbulent markets. *California Management Review, 42*(4), 24–54.

Schneider, B., Gunnarson, S. K., & Niles-Jolly, K. (1994). Creating the climate and culture of success. *Organizational Dynamics, 23,* 17–29.

Soliman, F., Spooner, K. (2000). Strategies for implementing knowledge management: role of human resources management. *Journal of Knowledge Management, 4*(4), 337–345.

Sun, P. Y.-T., & Scott, J. L. (2005). An investigation of barriers to knowledge transfer. *Journal of Knowledge Management, 9*(2), 75–90.

Sveiby, K. (2001). A knowledge-based theory of the firm to guide in strategy formulation. *Journal of Intellectual Capital, 2*(4), 344–358.

Sveiby, K., & Simons, R. (2002). Collaborative climate and effectiveness of knowledge work—an empirical study. *Journal of Knowledge Management, 6*(5), 420–433.

Szulanski, G. (1995). Unpacking stickiness: an empirical investigation of the barriers to transfer best practice inside the firm. In *Academy of Management Proceedings* (pp. 437–441).

Szulanski, G. (1996). Exploring internal stickiness: impediments to the transfer of best practice within the firm. *Strategic Management Journal, 17*, 27–43.
Teece, D., Pisano, G., & Shuen A. (1997). Dynamic capabilities and strategic fit. *Strategic Management Journal, 18*, 510–533.
Van Maanen, J., & Barley. S. R. (1984). Occupational communities: culture and control organizations. In B. M. Staw & L. L. Cummings (Eds.), *Research in organization behaviour* (Vol. 6, pp. 287–366). Greenwich, CT: JAI Press.
von Krogh, G. (1998). Care in knowledge creation. *California Management Review, 40*(3).
von Krogh, G., Nonaka, I., & Rechsteiner, L. (2012). Leadership in organizational knowledge creation: a review and framework. *Journal of Management Studies, 49*(1), 240–277.
Wah, L. (2000). Making knowledge stick. In J. W. Cortada, & J. A. Woods (Eds.), *The knowledge management yearbook 2000–2001* (pp. 145– 156). Woburn, MA: Butterworth-Heinemann.
Wang, C., & Hu, Q. (2020). Knowledge sharing in supply chain networks: effects of collaborative innovation activities and capability on innovation performance. *Technovation, 94,* 102010.
Wang, S., & Noe, R. A. (2010). Knowledge sharing: a review and directions for future research. *Human Resource Management Review, 20*, 115–131.
Wenger, E., McDermott, R., & Snyder, W. (2002). *Cultivating communities of practice: a guide to managing knowledge*. Harvard Business Press.
Yang, J. T. (2007). The impact of knowledge sharing on organizational learning and effectiveness. *Journal of Knowledge Management, 11*(2), 83–90.
Yang, J.-T. (2007). Knowledge sharing: investigating appropriate leadership roles and collaborative culture. *Tourism Management, 28*(2), 530–543.
Yang, H., & Wu, T. (2007). Knowledge sharing in an organization. *Technological Forecasting & Social Change.*

Knowledge Transfer: An Emerging Element of a Learning Organisation in Family Businesses

Nick Chandler

Abstract This chapter presents an extensive review of the literature for a number of key areas for examining the family business in the context of a learning organisation and how knowledge management practices in family businesses may be considered as an important indicator of central elements of a learning organisation. First, arguments for and against family businesses being seen as learning organisations are presented. The theoretical background defining what constitutes and what does constitute a learning organisation is considered for SMEs in general, and family businesses in particular. The types of knowledge are presented and assessed in the context of family businesses and the obstacles and drives towards knowledge transfer as part of a learning organisation. Knowledge management are then contemplated within the context of one crucial knowledge transfer long-term scenario: the succession process. The chapter is concluded with an assessment of the key findings of this review and considerations for future research directions.

1 Introduction

External environmental pressure s have been and are continually pushing for change. In a business context, the ability to change is pre-empted by the need to identify environmental changes and take advantage of them as a means towards both survival and growth in the longer term. In this way, recognising and adapting to change can be considered important, but there is also the need to see change as an ongoing process. Researchers have also found that this view of change is key to maintaining the edge over competitors (Kaufman 1992) and as a part of this Lucas et al. (1996) highlight that organisational learning plays an important role. With the combined aspects of continuous change and organisational learning, Nonaka et al. (2000, p. 6) see an organization as "an entity that creates knowledge continuously".

In a similar 'process mindset', becoming a learning organisation is also a continuous process and one in which the early beginnings may be seen in Small and

N. Chandler (✉)
Budapest Business School, University of Applied Sciences, Budapest, Hungary
e-mail: chandler.nicholas@uni-bge.hu

C. Machado and J. P. Davim (eds.), *Knowledge Management and Learning Organizations*,
https://doi.org/10.1007/978-3-030-71079-8_4

Medium-sized Enterprises (SMEs). Although much of the early studies of learning organisations have been applied to public sector organisations (Finger and Brand 1999), other researchers have supported the idea that not only SMEs but family businesses may be on the road to becoming learning organisations. Birdthistle and Fleming (2005) highlight that aspects such as flat structures, less bureaucracy and better communication may be more conducive for learning organisations than other business forms and sizes.

The learning aspect of business is of high importance in most industries and in family businesses and it is also of utmost value. Through learning key knowledge, companies can attain the much-needed competitive advantage (Harrison and Leitch 2005) and family businesses over time may also reach key specializations (Deakins 1998). According to Birdthistle and Fleming (2005), a study of family firms will offer a rich context in which to examine the learning organization. Moreover, with increased complexity, the learning organisation may be the means for tying together the array of elements that organisations are concerned with when analysing the internal and external environment (Jones and Hendry 1992).

There are certain contexts when knowledge management comes to the fore to a greater extent than in routine business life. Brannback et al. (2008) highlight that the succession process in family businesses is a "fundamental knowledge sharing, creation and renewal process". Organisational knowledge can increase competitiveness through family business succession through greater innovation, creativity, and competitiveness due to the knowledge transferred to the successor (the next generation) (Duh 2014; Nonaka and von Krogh 2009).

This chapter will consider the arguments for and against consideration of the family business as a learning organisation. From this, the knowledge management aspect will be considered in a family business by first considering the types of knowledge and then taking a 'process mindset' to knowledge management in learning organisation by examining relevant aspects of succession in family business such as knowledge transfer, phases of transfer and other relevant issues. The chapter is then concluded, along with some suggestions for further research.

2 Family Business as a Learning Organization

This chapter is concerned with a number of elements: the learning organisation, family businesses and succession. In this section, the arguments for and against a family business being considered as a learning organisation will be presented. First, however, the terms 'family business' and 'learning organisation' will be defined.

There has been some debate about what constitutes a family business caused by a plethora of varying definitions and a lack of consensus on a single definition. Handler (1989) identified four elements that constitute a family firm: degree of ownership and management by family members; interdependent sub-systems; generational transfer, and multiple conditions. Despite this, a number of works prior and post Handler's four dimensions focus on single elements, based upon relevance to particular studies.

Earlier works such as Barnes and Hershon (1976) define the family business with a focus on the ownership being in the hands of a single family. Likewise, later works such as Birdthistle (2003) defined a family business as a "proprietorship, partnership, corporation or any form of business association, which is classified as an SME and where the majority ownership is held by the family and family members are employed in the family business and/or the family is represented on the Board of Directors" (p. 76).

Senge (1990) referred to learning organisations as those that 'tap' employee's commitment and capacity to learn, regardless of their level in the organisation and organisations learn through the learning of individuals. The employee aspect was broken down by Yang et al. (2004) into individual and team levels. Whereas, Watkins and Marsick's (1993), who broke the employee aspect into team, individual and organisational levels, combined this human aspect of learning with that of structure. The focus on the human aspect of a learning organisation is often based upon the concept of a 'community of practice' which refers to a "group of people informally bound together by shared expertise and passion for a joint enterprise" (Wenger and Snyder 2000, p. 139). Watkins and Marsick (1993) view the human aspect or 'people level' as focussed on the individual and team. This level has four activities: to create continuous learning opportunities; to promote inquiry and dialogue; to encourage collaboration and team learning; and to empower people toward a collective vision. They also refer to another level, the 'structural level', and has three functions: to connect the organisation to its environment; to establish systems to capture and share learning; and to provide strategic leadership for learning. Watkins and Marsick (1993) assert that it is the combination of these two levels and their associated activities that culminate in the learning organisation.

Ortenblad (2002) put forward four perspectives of the learning organisation concept. The first perspective sees learning as the application of knowledge across and at different levels in the organisation, creating an 'organisational mind' that constitutes the learning organisation. The second perspective takes the human aspect referred to earlier in this section concerning the learning organisation on an individual level, as individuals learn during time spent at the workplace. For the third perspective, the learning organisation is seen as one that centres around expediting the learning of its employees and thus is a perspective of the organisational level. The fourth perspective is concerned with learning structure and sees the learning organisation as a flexible entity. Örtenblad's (2002) perspectives are also seen as types of learning organisation and are summarised as: organizational learning; learning at work; learning structure; and climate for learning.

For family businesses, there are a range of benefits for becoming a learning organisation on the organisational level, such as greater innovation, stronger financial performance (Slater & Narver 1995) and, as referring to in the introduction, more effective responses to the external environment (Harrison and Leitch 2005), which will all aid in achieving a stronger competitive advantage. On the team and individual level, the social groups of family and non-family employees will have an impact upon learning. The following paragraphs will consider further how family businesses (fbs) fit the perspectives of the learning organisation presented in this section.

If we consider the first perspective of Ortenblad (2002), the problem presents itself that learning organisation and organisational learning are often used interchangeably (Birdthistle 2008).

Two writers have sought to clarify this difference. Edmondson and Moingeon (1998) point out that it is organisational learning is distinguished as a human and organizational process, which is reinforced by Garavan (1997), who points out that the term 'organisational learning' describes and quantifies learning. This descriptive aspect indicates that 'organisational learning' indicates how learning in developed (Yeo 2005) and as this chapter examines the emergent aspects of the learning organisation, organisational learning itself will indicate how learning develops in the family business, through the example of succession.

In answer to the 'how' organisational learning develops in family businesses, Lubatkin et al. (2007) found that learning occurs through the characteristic informality of small firms, such as the majority of family businesses, which on the other hand, also may result in correspondingly less formal practice such as performance management systems to encourage learning (Kidwell et al. 2012).

The second perspective of Ortenblad (2002) views the learning organisation on an individual level and how learning at work takes place. There is a potential for barrier between family and non-family employees as an 'inner circle' develops of the family members (Eddleson and Kellermans 2007). Being privy to business information as a family member, regardless of position or status will also potential create barriers between those 'on the inside' and those 'on the outside' as well as invoking a feeling of not being trusted in the organisation. Another individual level human consideration is that of management style. Family businesses have a typically paternalistic style (Heidrich et al. 2016), which may lack the flexibility and autonomy for employees that are conducive to a learning organisation and aim to ensure control is kept within the family (Miller & Le Breton-Miller 2006).

Within the scope of the second perspective, the social capital aspects of family firms seem to encourage the emergence of a learning organisation, although it may be claimed that learning at work is limited to family members in many cases (Jack 2005). However, the liability of newness necessitates interactions with outsiders and potential alliances as the owner/entrepreneur learns the ropes (Yeung and Soh 2000).

The third perspective of the learning organisation (Ortenblad 2002) is concerned with enabling learning of all employees in the organisation. If we consider Birley et al.'s (1999) typology of family involvement, then the level of involvement and commitment of family members may indicate how on board they are likely to be in enabling learning in fbs. There are three types of firms. The first is family-in and in this case the family is very much involved in the daily running of the business and wanted family members to take over the business in time. As covered in a later section in this chapter, this desire for family member(s) to succeed prerequires the education of children, knowledge transfer and a longer-term view of the business. The second is family-out, where the family members are not pressured to take over the fb and so less support will come from the owners in terms of involvement and commitment. This may also mean that non-family employees may hold the top management positions. This does not mean that there is no opportunity for instating learning from non-family top management, although the lack of support from above may become an obstacle to any attempts to facilitate learning – especially if costs are high for this facilitation. The third category is referred to as 'family-business jugglers', as there is an attempt to achieve a balance between family and business issues. For this type, it is unclear if this need for balance will present itself as an obstacle or route towards enabling employees to learn.

The fourth perspective was concerned with encouraging a climate for learning and a degree of flexibility. In terms of a climate conducive to learning, it should not be forgotten that family businesses are created by entrepreneurs. This is important as learning is a key part of entrepreneurship, not only due to the liability of newness and attempts to overcome it, but also as entrepreneurial learning (EL) has the potential to enhance entrepreneurs' capabilities (Jiao et al. 2010). The centralized structure of family firms may allow the power of owner-managers to enable employees to learn (Aguilera & Crespi-Cladera 2012), but it could also act as an obstacle if knowledge is seen as predominantly held by the owner-manager and the characteristic hierarchical structure is less conducive to learning organisations, where a flat structure is preferred. This is further impacted if top-down communication is seen as the means by which knowledge is shared and may indicate a lack of flexibility. The climate for learning also is related to organisational culture. The aforementioned centralized control systems and paternalistic style have been argued as caused performance-avoidance cultures, and the associated cautiousness and risk-aversion may inhibit the activities of a learning organisation (Birdthistle 2003), especially as supportive leadership is claimed as crucial to developing a learning climate (Ortenblad 2013).

The possible emergence of a performance avoidance culture in family businesses does not mean that the culture is entire unsuited to a learning organisation. The values of long-term orientation through a view towards succession, harmony, commitment, trust and knowledge transfer (at least between family members) (Chandler et al. 2019) all may enhance the potential for developing a learning organisation. Although Chrisman et al. (2012) suggest external advice is valued less in fbs, the context of succession suggests a need for external sources of knowledge, which will be covered further in a later section of this chapter.

A model that integrates the many levels of a learning organisation is Crossan's 4i framework (Crossan et al. 1995). On an individual level, the employee focuses on intuition and interpretation. At the group level, integration is required to develop shared beliefs and behaviours, which ties in with the important role culture plays in the learning organisation. On the organisational level, the fourth 'i' stands for institutional learning that emerges as a result of organisational systems in place. When these levels are considered, then the challenges facing family businesses in becoming a learning organisation are substantial.

The challenges facing family businesses in becoming a learning organisation go beyond the informality and lack of formal systems. Choueke and Armstorng (1998) highlight that sharing power and decision making is one such significant challenge. They indicate that this unwillingness to share may stem from yet another hurdle: short-termism.

Despite the challenges facing family businesses, some fbs go against the stereotypical elements of family businesses that are listed as obstacles to becoming learning organisations. Shepherd (2016) is an example of how family businesses are open to and encouraging organisational learning. Shepherd (2016) cites the example of Barleyfields, a family-run business in operation for 10 years. The firm embraced Critical Reflection Action Learning as an approach for commencing an organisational change program.

The above example may be seen as displaying only one of the elements that were referred to earlier as signs of a learning organisation. A family business may not have the resources to take on all four perspectives of the learning organisation in one go. Birdthistle and Fleming (2005) found in a study of SMEs in Ireland that it is rare for these smaller firms to exhibit all the characteristics of a learning organisation. Some family businesses were found to be challenged by the need to create continuous learning opportunities and stick to informal learning. On the positive side, the study found that some family businesses did have the systems for facilitating learning, such as a training budget and personal growth objectives. Furthermore, they found family businesses also had communication systems that encouraged open and honest feedback with other employees, and thereby developed a 'listening culture'.

On the group level, Birdthistle and Fleming (2005) found that some family firms encouraged collaboration and team learning, but as with the other results, a lot depended on the size of the family firms as micro family firms lacked in soe of these areas, when compared to small and medium-sized family firms.

These findings illustrate well that family firms should not be 'tarred with the same brush' as some have started to develop as learning organisations and becoming a learning organisation should be seen as a continual process rather than a fixed state or list of criteria to be met. Burgoyne (1995) highlights that a learning organisation is not a standard formula that organisations need to follow to achieve 'learning organisation status'. From a family business perspective, Cunningham et al. (2016) highlight a number of studies that vary greatly in their approach to sharing knowledge, with some sharing and encouraging collaborative experimentation, whilst other keep knowledge firmly in the hands of the fb founder. Furthermore, the four perspectives that were put forward by Ortenblad (2004) are also confirmed by the same author as not all being present in organisations, which have resulted in the term 'partially learning organisation'.

In sum, it can be seen that family businesses have potential to be considered as learning organisations or having elements of the learning organisation emerging as it grows over time towards a fully-fledged learning organisation. The following section will consider the types of knowledge in family businesses for considering this aspect of learning further, before the knowledge management and transfer involved in succession are presented in the latter sections of this chapter.

3 Types of Knowledge in Family Businesses

In this chapter we will consider the role of knowledge transfer in the succession process of family businesses as an indicator of the emergence of the learning organisation. This section will present the types of knowledge and how they fit into the picture of the family business as a learning organisation.

For all the different types of knowledge, some are more easily transferred to other members of the organisation than others. General knowledge is an example of explicit knowledge and so is easy to transfer. Technical knowledge, on the other hand, can only be transferred if the receiver/learner has a suitable background to understand the rudiments of the technical aspects (Royer et al. 2008).

The harder types of knowledge for transfer is implicit or tacit knowledge, such as intuitive knowledge, based upon training and experience (Boyd et al. 2015). An earlier work by Boyd and Royer (2012) specify three types of experiential knowledge. The first is idiosyncratic knowledge. This is detailed knowledge specific to the given situation such as location e.g. the local environment. The second is subject-related experiential knowledge. This may require a specific set of skills on how to use certain tools or materials before knowledge can be transferred (see Patriotta 2007). The third is network-related experiential knowledge—as mentioned earlier in this chapter, the need to overcome the liability of newness and the strengths of social capital indicate that this knowledge may be crucial to the survival, both short- and long-term, for family businesses.

Regardless of the type of tacit knowledge, it can be seen that this form of knowledge that is difficult to transfer as some are based upon ideas or beliefs that are difficult to express and others rely on the existing know-how or ability before knowledge can be transferred (Cabrera-Suárez et al. 2001). In these cases, Ortenblad's (2002) second perspective of the learning organisation that is 'learning at work' applies, as the most suitable method for transfer is 'learning by doing'. In the case of family businesses, much of the tacit knowledge is initially in the mind of the founder.

Explicit knowledge can be expressed in a more formal, systematic way than tacit knowledge (Nonaka et al. 2000) and the example of general knowledge referred to earlier as an easier knowledge type to transfer is often also the type of knowledge that is least relevant for a family business to achieve competitive advantage, such as numerical ability (Boyd and Royer 2012). It should be noted here that tacit and explicit knowledge are not mutually exclusive, but rather interact with one another on a group and individual level (Nonaka and von Krogh 2009).

As mentioned earlier in this chapter, family businesses have strong social capital. From a knowledge perspective, strong social capital enables family firms to develop tacit knowledge (Cabrera-Suarez et al. 2001). This is due primarily to high levels of mutual trust and understanding between family members involved with the business. From our model of perspectives of the learning organisation, this indicates relevance to the second perspective of the learning organisation, i.e. on the group and individual level, and with consideration for the development of social capital. However, there is some debate in this area as McAdam and Reid (2001) argue that small businesses are less likely to rely on social interaction than larger businesses. Despite the difference between large and small businesses, SMEs, including family businesses, utilize shared networks and relationships in their day-to-day routines (Lin 2013).

In addition to the type of knowledge, there are a number of elements that may act as enablers or obstacles to the transfer of knowledge, due to the nature of family businesses. Durst and Wilhelm (2012) highlight the centrality of the owner-manager, resulting in the close attention and control of daily operations, and that this combined with a lack of resources and expertise are likely to culminate in the retention of knowledge in the mind of the owner. Conversely, Desouza and Awazu (2006) argue that small businesses are prone to knowledge transfer precisely because of their small size, with less hierarchy and close contact between all employees, although they also argue that the centralised storage of knowledge in the mind of the owner is likely to result in the higher likelihood of the lo ss of knowledge. As referred to earlier in this chapter, knowledge management in most small businesses occurs in an informal way, although the concept and vocabulary used in knowledge management is also employed (Hutchinson and Quintas 2008).

4 Knowledge Transfer and Succession

In the previous section we have considered the complexity of types of knowledge and their potential transfer in family businesses, and this complexity is emphasised by pivotal writers in the field (e.g. Grant 1996; Szulanski (1996). To uncover the hidden complexities of knowledge transfer in small businesses, Le Breton-Miller et al. (2004) suggest the topic of succession in family businesses as a suitable focus and highlight how knowledge is transferred to the successor from the founder from an early age, and transfer continues over time. Malinen (2004) found that the transfer of knowledge was ranked the third most important problem of family business succession, after the issues of finding a suitable successor and taxation.

The theoretical framework for knowledge transfer is related to knowledge creation theory (Nonaka and von Krogh 2009). To increase knowledge capacity demands the use and development of both explicit and tacit types of knowledge amongst employees. On an organisational level, knowledge creation is therefore connecting this knowledge created by employees to the organisation's knowledge system (Nonaka et al. 2006).

As the knowledge is often retained in the mind of the founder (incumbent) in family businesses, the onus is on the incumbent to transfer knowledge (Cadieux 2007), although knowledge can come from other areas in the organisation and outside of it. There are various areas that may facilitate the transfer of knowledge such as a shared vision, strong relationships and how knowledge is treated (Wasim et al. 2018).

In the context of knowledge transfer, Boyd et al. (2015) highlight the specificities of family businesses. First, they assert that family firms have an easier job of transferring knowledge to the successors, when compared to other firms, as successors were nurtured and introduced to the business' networks from a young age. Sambrook

(2005) gives three different types of knowledge that should be transferred to successors in family businesses: technical knowledge (K1); knowledge about the firm and how it works (K2); and tacit knowledge, which depends to a great extent on the willingness of the incumbent to transfer knowledge and the successor to receive it (K3) (see also Surdej and Brzozowski 2017). It is K3 that Sambrook (2005) suggests is the most critical to transfer in the succession process, and some studies suggest that this type of knowledge should remain within the family member (Bjuggren and Sund 2002).

To facilitate knowledge transfer, the incumbent has the duty of showing enthusiasm and pride when the successor is successful in family business operations as well as open to new ways of managing the business, whereas successors need to appreciate the work of the incumbent in transferring knowledge and the value of the existing systems set up by the incumbent (Cabrera-Suárez et al. 2001; Royer et al. 2008). These aspects raise the importance of the relationship between the incumbent and successor in knowledge management in family firms (Chirico and Salvato 2008). Moreover, Hadjielias et al. (2010) contend that the transfer of knowledge in general and entrepreneurial orientation in particular during family business successful are crucial to the long-term survival of the firm.

From a learning and knowledge perspective, we have seen in this section that the incumbent has to amass knowledge to meet the demands of running a business and require the founder to learn about the business environment as well, as a means of ensuring long-term growth (Cope 2005), which serve to highlight that family firms in fact have a longer term view despite often being labelled as having a homogenous 'short-termist approach' in many circles. Boyd et al. (2014) also support the long-termism is incumbents in family businesses, as well as the need for external sources of knowledge and for networking, or building social capital, through the training of the successor with external experience and education, which also demands the motivation and need for building and maintaining relationships for external sources of knowledge and learning (see also Tapics and Fernandéz Moya 2012).

As a part of this longer-term perspective, succession is viewed as an on-going process of preparing the successor for taking on the family business from the incumbent. Sharma and Srinivas Rao (2000) conducted a comparative study of two countries (Canada and India) and found a number of traits that incumbents rated as signs of the preparedness of successors. In India, incumbents rated the blood and family relationships above all else, whereas in Canada, the focus was on interpersonal skills, previous performance and experience gained. Sharma and Srinivas Rao (2000) also found a number of attributes that that successors should possess, whether in Canada or India: integrity and commitment to the business. Age, birth order, and gender of the successors was unimportant for all incumbents in the study.

As a final note on the transfer of knowledge between generations in family businesses, research has found different attitudes to succession between first-, second and third-generation family businesses, which may also indicate that as generations pass, knowledge transfer and management will also change (Brannback et al. 2008) through the creation of new traditions between family members as a result of the way in which knowledge is managed and transferred. To consider this process of transferring knowledge (Buckman et al. 2019) and the differences between generations further, the following section will look at the stages of knowledge management that emerge during the succession process.

5 Stages of Knowledge Management in Succession

Cater and Kidwell (2014. p. 1) define succession as a "dynamic process involving the transfer of both the management and ownership of a family firm to the next generation". As the process of transferring tacit knowledge hinges on experience, it is close to the theoretical framework of Kolb et al. (2001) that learning is experiential and relies on a range of components such as incumbents, successors, the family, the type of business, and the stakeholders (Csizmadia et al. 2016).

Before considering the stages of knowledge transfer in succession, the stages of the succession process form the basis for this process. Aronoff and Ward (2011) identified the stages of the process. First, they name the 'development phase'. This is where the incumbent passes on the necessary skills and knowledge. The next phase is the 'transitional phase' where the incumbent prepares the successor specifically to lead the family business. Finally, the 'takeover phase' is where authority and control for decision making is passed from the incumbent to the successor. In contrast with this Cater and Kidwell (2014) put forward a four-stage model for succession. They add a 'bridging phase' where the successor has taken over the leadership position but the incumbent works in partnership with the incumbent to firmly establish the successor's leadership position before fully standing down. There is some debate on this stage as Le Breton-Miller et al. (2004) put forward that the successor is often left to their own resources, perhaps as a display of standing on their own two feet, rather than have a 'bridging phase'.

The stages of knowledge management in succession are often described in the long-term plans for succession. There is some debate as to when the succession plan should begin. According to Ibrahim et al. (2001) the succession plan begins when the successor begins work in the family business, however – as presented earlier in this chapter – the transfer of knowledge as a part of the succession process often takes place from the early years and long before entering the business for work (see Cater and Kidwell 2014).

When comparing two family businesses, Boyd et al. (2015) found that the first type of knowledge to be transferred from the incumbent to the successor was idiosyncratic knowledge, which occurred in the early stages of the succession process, followed by network-related knowledge. However, Boyd et al. (2015) also found that the industrial and institutional context affect the knowledge types transferred and the method of knowledge transfer.

Pham et al. (2019) studied the father-son dynamic within the context of stages of succession. They found a range of roles that the father undertakes during the succession process, which ay give insight into the stages of knowledge transfer. First the father acts as an example to the successor. This is during the successor's childhood (Haldin-Herrgard 2007) and may be the implicit, tacit, or general knowledge as values, beliefs and perceptions are pass on during the course of the family relationship (K3). In later years, the father takes on the role of a supporter as the incumbent pushes the successor to gain external knowledge about business in the form of formal education (Bozer et al. 2017) and work experience at companies (Sardeshmukh and Corbett 2012; Duh and Letonja 2013). This appears to be the general knowledge in the form of education and experience which will aid not only in running the family business through remedying any deficiencies in the incumbent's knowledge and skills (Man et al. 2016), but also the acquired knowledge acts as a necessary foundation for the later transfer of technical knowledge. It should be noted, however, that formal education may also decrease the desire of the successor to take over the family business (Palliam et al. 2011). Following this phase, as the successor has gained sufficient experience (in the eyes of the incumbent), they join the family business. At this stage, the incumbent takes on the role of mentor and trouble-shooter with the successor as they handle day-to-day operations (see also Sabri et al. 2016). Since the successor has gained experience and qualifications, this is the likely stage for the transfer of any technical knowledge in the family business (K1). Once the incumbent has passed the reins to the successor, the successor takes a role as an advisor. Throughout all these stages we may see the transfer of knowledge of the firm and how it works (K2), although to a lesser extent when the successor is acquiring knowledge from external sources.

Finally, Buckman et al. (2019) developed a model of the succession process which combines many of the elements presented in this section: the stages of the succession process, the knowledge inputs and the family factors that may influence the succession process either as drives or obstacles. Based on this and a further study by Chandler, Mosolygó-Kiss, & Heidrich (2019) of knowledge transfer in the succession process, the following model is put forward:

6 Conclusions and Future Directions

This chapter presents the argument that a family business can be seen as a learning organisation. According to the literature, the family business has some elements that relate to the four perspectives of a learning organisation, but not all of them. However, most organisations fall into the category of a partial learning organisation.

Empirical studies often stress the importance of effective succession for the long-term success of the family business, but recent studies also focus on the succession process. The knowledge transfer element has been considered to some extent in terms of the likely types of knowledge and the stages in the life of the successor and incumbent as knowledge is transferred. There still remains work to be done on the disparities between countries as existing studies indicate the potential for national differences.

This review has contributed to existing knowledge by summarizing the key studies and developments in this field. The summation of many of the issues can be seen in the model in Fig. 1. However, there is much scope for further development as this model does not indicate empirical evidence of the actual knowledge types transferred during the succession process.

The incumbent-successor dynamic presented here indicates the need for a dyadic approach to research that will compare and contrast these two potentially at odds, perspectives. The detail required to develop this further also points towards a qualitative approach with interviews with both generations, and future generations, if possible.

Becoming a learning organisation and ensuring effective succession are clearly not mutually exclusive, but rather one can help the instigation of the other. Furthermore, these two elements will contribute to the longevity and stable growth of the family business.

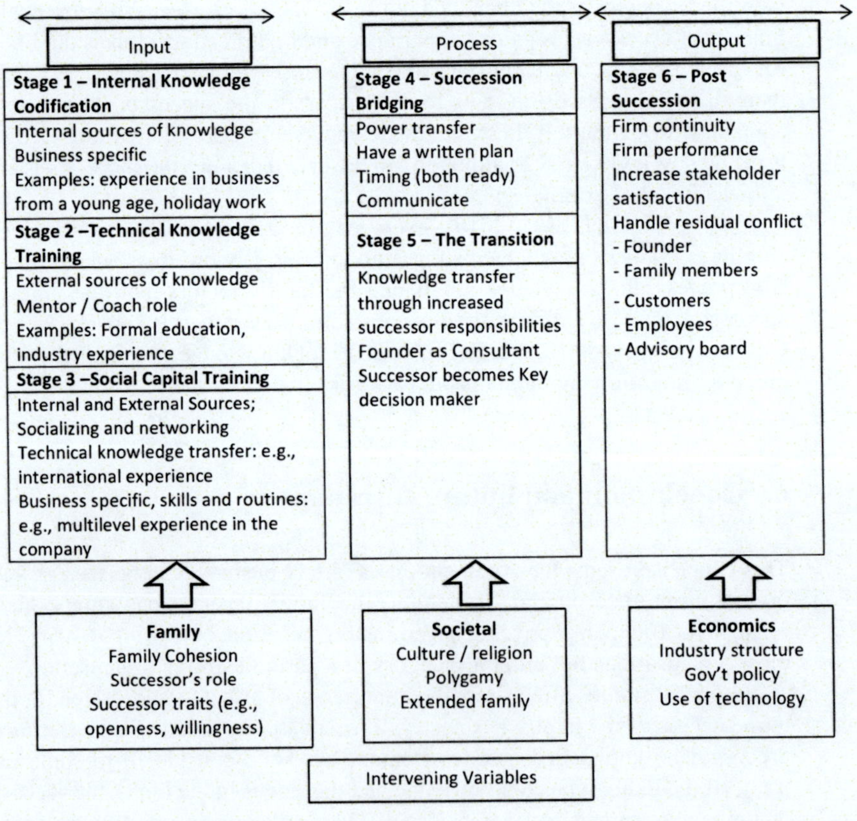

Fig. 1 The succession process and knowledge transfer. *Source* Adapted from Buckman et al. (2019) and Chandler et al. (2019)

References

Aguilera, R. V., & Crespi-Cladera, R. (2012). Firm family firms: current debates of corporate governance in family firms. *Journal of Family Business Strategy, 3*(2), 63–69.

Aronoff, C. E., & Ward, J. L. (2011). *Make change your family business tradition*. New York, NY: Palgrave MacMillan.

Barnes, L. B., & Hershon, S. A. (1976). Transferring power in the family business. *Harvard Business Review, 53*(4), 105–114.

Birdthistle, N. (2003) Educating the family business: an investigation into centres of excellence for family businesses and family business educational programs. In *SMEs in the Knowledge Economy: Proceedings of the 26th ISBA Conference*, 12–13 November (pp. 76–96). Institute for Small Business Affairs.

Birdthistle, N. (2008). Family SMEs in Ireland as learning organizations. *The Learning Organization, 15*(5), 421–436.

Birdthistle, N., & Fleming, P. (2005). Creating a learning organisation within the family business: an Irish perspective. *Journal of European Industrial Training, 29*(9), 730–750.

Birley, S., Ng, D., & Godfrey, A. (1999). The family and the business. *Long Range Planning, 32*(6), 598–608.

Bjuggren, P. O., & Sund, L. G. (2002). A transaction cost rationale for transition of the firm within the family. *Small Business Economics, 19*(2), 123–133.

Boyd, B., & Royer, S. (2012). The suitability of internal versus external successors: Relevant knowledge types in family business succession. *International Journal of Management Practice, Inderscience Enterprises Ltd, 5*(4), 361–382.

Boyd, B., Royer, S., Pei, R., & Zhang, X. R. (2014). Knowledge Transfer in Family Business Successions: Comparing a Chinese and European Family Business. In *The International Family Enterprise Research Academy 2014 Annual Conference: Co-operation Within and Amongst Family Businesses.*

Boyd, B., Royer, S., Pei, R., & Zhang, X. R. (2015). Knowledge transfer in family business successions: Implications of knowledge types and transaction atmospheres. *Journal of Family Business Management, 5*(1), 17–37.

Bozer, G., Levin, L., & Santora, J. (2017). Succession in family business: Multi-source perspectives. *Journal of Small Business and Enterprise Development, 24*(4), 753–774.

Brannback, M., Carsrud, A., & Schulte, W. (2008). Exploring the role of Ba in family business context. *VINE, 38*(1), 104–117.

Buckman, J., Jones, P., & Buame, S. (2019). Passing on the baton. *Journal of Entrepreneurship in Emerging Economies.*

Burgoyne, J. G. (1995). Learning from experience. *Personnel Review* (September 1).

Cabrera-Suárez, K., De Saá-Pérez, P., & Garcia-Almeida, D. (2001). The succession process from a resource and knowledge-based view of the firm. *Family Business Review, 14*(1), 37–48.

Cadieux, L. (2007). Succession in small and medium-sized family businesses: Toward a typology of predecessor roles during and after instantement of the successor. *Family Business Review, 20*(2), 95–109.

Cater, J. J., & Kidwell, R. E. (2014). Function, governance and trust in successor leadership groups in family firms. *Journal of Family Business Strategy, 5*(3), 217–228.

Chandler, N., Mosolygó-Kiss, A., & Heidrich, B. (2019). Transferring responsible leadership: The manifestation of responsible leadership characteristics during family business succession. *Academy of Management Global Proceedings, Slovenia,*. https://doi.org/10.5465/amgblproc.slovenia.2019.0144.abs.

Chirico, F., & Salvato, C. (2008). Knowledge integration and dynamic organizational adaptation in family firms. *Family Business Review, 21*(2), 169–181.

Choueke, R., & Armstrong, R. (1998). The learning organisation in small and medium-sized enterprises: A destination or a journey? *International Journal of Entrepreneurial Behavior & Research, 4*(2), 129–140.

Chrisman, J. J., Chua, J. H., Pearson, A. W., & Barnett, T. (2012). Family involvement, family influence, and family- centered non- economic goals in small firms. *Entrepreneurship Theory and Practice, 36*(2), 267–293.
Cope, J. (2005). Toward a dynamic learning perspective of entrepreneurship. *Entrepreneurship Theory and Practice, 29*(4), 373–397.
Crossan, M., Lane, H., White, R. E., & Djurfeldt, L. (1995). Organizational learning: Dimensions for a theory. *International Journal of Organizational Analysis, 3*(4), 337–360.
Csizmadia, P., Mako, C., & Heidrich, B. (2016). Managing succession and knowledge transfer in family businesses: Lessons from a comparative research. *Vezetéstudomány/Budapest Management Review, 47*(11), 59–69.
Cunningham, J., Seaman, C., & McGuire, D. (2016). Knowledge sharing in small family firms: A leadership perspective. *Journal of Family Business Strategy, 7*(1), 4–46.
Deakins, D. (1998). Entrepreneurial learning and the growth process in SMEs. *The Learning Organization, 5*(3), 144–155.
Desouza, K. C., & Awazu, Y. (2006). Knowledge management at SMEs: five peculiarities'. *Journal of Knowledge Management, 10*(1), 32–43.
Duh, M. (2014). Family business succession as knowledge creation process. *Kybernetes, 43*(5), 699–714.
Duh, M., & Letonja, M. (2013). Knowledge creation in family businesses and its importance for building and sustaining competitive advantage during and after succession. *International Journal of Business Administration, 4*(6), 120–133.
Durst, S., & Wilhelm, S. (2012). Knowledge management and succession planning in SMEs. *Journal of Knowledge Management, 16*(4), 637–649.
Eddleston, K. A., & Kellermans, F. W. (2007). Destructive and productive family relationships: A stewardship theory perspective. *Journal of Business Venturing, 22*(4), 545–565.
Edmondson, A., & Moingeon, B. (1998). From organizational learning to the learning organization. *Management learning, 29*(1), 5–20.
Finger, M., & Brand, S. B. (1999). The learning organisation in the public sector. In M. Easterby-Smith, J. Burgoyne, & L. Araujo (Eds.), *Organisational learning and the learning organisation; Developments in theory and practice* (pp. 130–156). London: Sage Publications.
Garavan, T. (1997). The learning organization: A review and evaluation. *The Learning Organization.*
Grant, R. M. (1996). Towards a knowledge-based theory of the firm. *Strategic Management Journal, 17*(S2), 109–122.
Hadjielias, E., Hamilton, E., & Howorth, C. (2010). *Entrepreneurial learning in the family management group: An organisational learning perspective*. Centre for Family Business Lancaster University Management School.
Haldin-Herrgard, P. T. (2007). *Tacit knowledge diffusion in family business succession, Presented at the 3rd Workshop on Family Firms Management Research, June 3rd–7th.* Sweden: Jonkoping.
Handler, W. C. (1989). Methodological issues and considerations in studying family businesses. *Family Business Review, 2*(3), 257–276.
Harrison, R. T., & Leitch, C. M. (2005). Entrepreneurial learning: Researching the interface between learning and the entrepreneurial context. *Entrepreneurship Theory and Practice, 29*(4), 351–371.
Heidrich, B., Németh, K., & Chandler, N. (2016). Running in the family—Paternalism and familiness in the development of family businesses. *Vezetéstudomány-Budapest Management Review, 47*(11), 70–82.
Hutchinson, V., & Quintas, P. (2008). Do SMEs do knowledge management? Or simply manage what they know?'. *International Small Business Journal, 26*(2), 131–154.
Ibrahim, A. B., Soufani, K., & Lam, J. (2001). A study of succession in a family firm. *Family Business Review, 14*(3), 245–258.
Jack, S. L. (2005). The role, use and activation of strong and weak network ties: A qualitative analysis. *Journal of Management Studies, 42*(6), 1233–1259.

Jiao, H., Ogilvie, D., & Cui, Y. (2010). An empirical study of mechanisms to enhance entrepreneurs' capabilities through entrepreneurial learning in an emerging market. *Journal of Chinese Entrepreneurship, 2*(2), 196–217.

Jones, A. M., & Hendry, C. (1992). *The learning organization: a review of literature and practice*. UK: Centre for Corporate Strategy and Change, University of Warwick.

Kaufman, R. S. (1992). *Why operations improvements programmes fail: Four managerial contradictions* (pp. 17–36). Fall: Sloan Management Review.

Kidwell, R. E., Hoy, F., & Ibarreche, S. (2012). "Ethnic" family business or just family business? Human resource practices in the ethnic family firm. *Journal of Family Business Strategy, 3*(1), 12–17.

Kolb, D. A., Boyatzis, R., & Mainemelis, C. (2001). Experiential learning theory: Previous research and new directions. *Perspectives on Thinking, Learning, and Cognitive Styles, 1*, 227–247.

Le Breton-Miller, I., Miller, D., & Steier, L. P. (2004). Toward an integrative model of effective FOB succession. *Entrepreneurship Theory and Practice, 28*(4), 305–328.

Lin, W. (2013). Research on knowledge sharing and interpersonal relationships: Empirical study of family firms and non-family firms. *Quality & Quantity, 47*, 151–166.

Lubatkin, M., Lane, P., Collin, S., & Very, P. (2007). An embeddedness framing of governance and opportunism: Towards a cross-nationally accommodating theory of agency. *Journal of Organizational Behavior, 28*(1), 43–58.

Lucas, B., Hult, G. T., & Ferrell, O. C. (1996). A theoretical perspective of the antecedents and consequences of organisational learning in marketing channels. *Journal of Business Research, 36*(3), 233–244.

Malinen, P. (2004). Problems in transfer of business experienced by finnish entrepreneurs. *Journal of Small Business and Enterprise Development, 11*(1), 130–139.

Man, T. W. Y., Mustafa, M., & Fang, Y. (2016). Succession in Chinese family enterprises: The influence of cognitive, regulatory and normative factors. *International Journal of Management Practice, 9*(4), 412–432.

McAdam, R., & Reid, R. (2001). SME and large organisation perceptions of knowledge management: Comparisons and contrasts'. *Journal of Knowledge Management, 5*(3), 231–241.

Miller, D., & Le Breton-Miller, I. (2006). Family governance and firm performance: Agency, stewardship, and capabilities. *Family Business Review, 21*(1), 73–87.

Nonaka, I., & von Krogh, G. (2009). Tacit knowledge and knowledge conversion: Controversy and advancement in organizational knowledge creation theory. *Organization Science, 20*(3), 635–652.

Nonaka, I., Toyama, R., & Konno, N. (2000). SECI, Ba and leadership: A unified model of dynamic knowledge creation. *Long Range Planning: International Journal of Strategic Management, 33*(1), 5–34.

Nonaka, I., von Krogh, G., & Voelpel, S. (2006). Organizational knowledge creating theory: Evolutionary paths and future advances. *Organization Studies, 27*(8), 1179–1208.

Örtenblad, A. (2004). The learning organization: Towards an integrated model. *The Learning Organization*.

Örtenblad, A. (2002). A typology of the idea of learning organization. *Management Learning, 33*(2), 213–230.

Örtenblad, A. (Ed.). (2013). *Handbook of research on the learning organization: Adaptation and context*. Cheltenham, UK: Edward Elgar.

Palliam, R., Cader, H., & Chiemeke, C. (2011). Succession issues among family entrepreneurs in countries of the Gulf. *International Journal of Business Administration, 2*(2), 25–34.

Patriotta, G. (2007) *Organizational knowledge in the making—How firms create, use, and institutionalize knowledge*. Oxford Scholarship Online.

Pham, T. T., Bell, R., & Newton, D. (2019). The father's role in supporting the son's business knowledge development process in Vietnamese family businesses. *Journal of Entrepreneurship in Emerging Economies*.

Royer, S., Simons, R., Boyd, B., & Rafferty, A. (2008). Promoting family: A contingency model of family business succession. *Family Business Review, 21*(1), 15–30.

Sabri, S. M., Haron, H., Jamil, N., & Ibrahim, E. N. M. (2016). Intergenerational knowledge transfer strategy framework for family firm. In M. A. Abdullah, W. K. Yahya, N. Ramli, S. R. Mohamed, & B. E. Ahmad (Eds.), *Regional conference on science, Technology and Social Sciences (RCSTSS 2014)* (pp. 957–969). Singapore: Springer.

Sambrook, S. (2005). Exploring succession planning in small, growing firms. *Journal of Small Business and Enterprise Development*.

Sardeshmukh, S. R., & Corbett, A. C. (2012). The duality of internal and external development of successors: Opportunity recognition in family firms. *Family Business Review, 24*(2), 111–125.

Senge, P. M. (1990). *The fifth discipline. The art and practice of the learning organization*. London: Random House.

Sharma, P., & Srinivas Rao, A. (2000). Successor attributes in Indian and Canadian family firms: A comparative study. *Family Business Review, 13*(4), 313–330.

Shepherd, G. (2016). How a small family run business adopted Critical Reflection Action Learning using hand drawn images to initiate organisational change. *Action Learning: Research and Practice, 13*(1), 69–78.

Slater, S. F., & Narver, J. C. (1995) Marketing orientation and the learning organization, *Journal of Marketing*, 59, July, 63–74.

Surdej, A., & Brzozowski, J. (2017). Assessing the readiness to family firm succession among CEE students. *Przedsiebiorczosc i Zarzadzanie, 18*, 11–22.

Szulanski, G. (1996). Exploring internal stickiness: Impediments to the transfer of best practice within the firm. *Strategic Management Journal, 17*, 27–43.

Tapies, J., & Fernandéz Moya, M. (2012). Values and longevity in family business: Evidence from a cross-cultural analysis. *Journal of Family Business Management, 2*(2), 130–146.

Wasim, J., Cunningham, J., Maxwell-Cole, A., & Taylor, J. R. (2018). Nonfamily knowledge during family business succession: A cultural understanding. *International Journal of Entrepreneurial Behavior & Research, 26*(1), 141–157.

Watkins, K. E., & Marsick, V. J. (1993). *Sculpting the learning organisation: Lessons in the art and science of systemic change*. San Francisco, CA: Jossey-Bass.

Wenger, E., & Snyder, W. M. (2000) Communities of practice: the organizational frontier, *Harvard Business Review*, January-February, pp. 139–145.

Yang, B., Watkins, K. E., & Marsick, V. J. (2004). The construct of the learning organisation: Dimensions, measurement and validation. *Human Resource Development Quarterly, 15*(1), 31–55.

Yeo, R. K. (2005). Revisiting the roots of learning organization: A synthesis of the learning organization literature. *The Learning Organization, 12*(4), 368–382.

Yeung, H., & Soh, T. M. (2000). Corporate governance and the global reach of Chinese family firms in Singapore. *Seoul Journal of Economics, 13*(3), 301–334.

Corporate Memory System: Key for Experienced Based Management

Yasemin Sen

Abstract Knowledge is the most valuable asset of an organization and this knowledge is mostly come from past experiences. In order to ensure a successful experienced based management of organizations, a well structured corporate memory system is important. For this reason, in this chapter theoretical basis of corporate/organizational memory is presented and based on the previous models of organizational memory, a new model of corporate memory system with a socio-technical approach is proposed.

1 Introduction

It is possible to say that corporate/organizational memory, which expresses the knowledge of organizations, has a past almost as old as the history of business. Every organization carries organizational memory in different places within the organization, whether in its individuals, culture or practices. Organizational memory, whose importance has been noticed since the 1900s, is now considered a valuable asset of organizations and the good management of this asset is important for the success of organizations. Organizations can manage their knowledge and use them in their current activities with a structure and system that they can create. Therefore, in order to turn valuable knowledge of organizations into effective operations, a clear understanding of corporate memory system is needed. In this chapter first of all definition, typology and importance of corporate memory system have been explained. After that, debates on corporate memory concept, basic models of corporate memory and new corporate memory system model topics have been addressed.

Y. Sen (✉)
School of Business, Istanbul University, Istanbul, Turkey
e-mail: ysen@istanbul.edu.tr

C. Machado and J. P. Davim (eds.), *Knowledge Management and Learning Organizations*,
https://doi.org/10.1007/978-3-030-71079-8_5

2 Definition of Corporate Memory

The concept of organizational/corporate memory has been discussed for more than a quarter of a century (Van Heijst et al. 1997) and there is an ambiguity regarding the definition of the concept (Ackerman 1994). The concept was tried to be explained based on individual memory and some researchers used the concept only as a metaphor (Argyris and Schön 1978), while others suggested that organizations also have the ability to think (Hedberg 1981). In addition, although it has been examined as content with an organizational memory storage perspective earlier (Walsh and Ungson 1991), it has started to be addressed as a system in the following years (Stein and Zwass 1995). The differences in definitions are actually due to the fact that organizational memory is handled by researchers from different levels and different disciplines (Lehner 2004). The most important issue to be considered in this regard is to clarify the goal and draw the boundaries of the concept well.

Regardless of the perspective, the starting point of organizational memory is the concept of individual memory. Memory, in its most general definition, is expressed as the ability to keep and retrieve past events (Martin Corbett 1997). In a definition made for memory, it is stated that the concept is the process or power of reproducing or recalling what is learned or acquired through associative mechanisms. In another definition, memory is expressed as the place where information is stored and retrieved, and what is known dynamically is changed (Croasdell et al. 1997). As can be seen from these definitions, even at the individual level, it has been handled in various ways such as memory, ability, process or information storage. Another definition that we can say includes all these elements is made by Lehner. Lehner defined memory as a system capable of storing perceived, experienced or lived events beyond their actual occurrence time period and bringing them back at a later time, in other words, as a system with this ability (Lehner 2004).

The definitions made at the organizational level have also been in this direction. In a basic sense, organizational memory is a general concept used to describe the storage, symbolization and sharing of corporate knowledge (Croasdell et al. 1997). It is seen that the focus is on the process rather than the content of the memory. Similarly, Tuomi also considered organizational memory as a process in which the past affects today and stated that within this time period it is necessary not only to focus on the buffer memories that mediate this effect, but also to consider the process itself (Tuomi 2000). Another definition that focuses on the process was made by Gaele Simon. Although Simon does not use the concept of "organizational memory" in his definition, he describes the organizational memory process. Simon has called this process as "knowledge capitalization" and described it as enabling the reuse of previously stored and modeled relevant knowledge of a particular domain in order to perform new tasks (Abecker et al. 1998).

Walsh and Ungson defined organizational memory as stored information retrieved from an organization's past for use in the decision-making process. Although the researchers also mention organizational memory processes in their study, as can be understood from the definition, their main focus has been the storage of information

(Walsh and Ungson 1991). According to Argyris and Schön, organizational memory is the map of the organization's past. The information constituting the organizational memory is scattered in individuals' minds, files, documents or computer memories, and this scattered information should be organized and translated into a whole picture for use in business activities and learning (Argyris and Schön 1978). Similarly, in this definition also, the carriers in which organizational information takes place are mentioned. Wegner et al. also built the definition on the knowledge repository approach and accepted the organizational memory as the level of collecting, coding, storing and updating the relevant information stock, which is ready for distribution, coded with the help of information technologies (Wegner et al. 1991). According to Hedberg, organizational memory constitutes the cognitive structures of information processing, the theory of action for the entire organization (Hedberg 1981).

Stein and Zwass, on the other hand, took the organizational memory from a system perspective and defined the concept as a system that functions as a means of using past knowledge in business activities and can result in an increasing level of efficiency for the organization (Stein and Zwass 1995). In this definition, unlike the others, the contribution of the organizational memory system to increase efficiency is emphasized (Lehner 2004). Corbett also took the organizational memory from a system perspective and suggested that the organizational memory system consists of interrelated elements that are constantly collected and recollected, constructed and reconstructed by human or non-human actors (Martin Corbett 1997).

From a broader perspective, the concept of organizational memory includes the technical, functional and social aspects of the job, employee and workplace, and includes elements that can be expressed in written records such as corporate manuals, databases, filing systems (Croasdell et al. 1997).

As a result, organizational memory is a concept that has both social and technical aspects and it would be more appropriate to handle it from a system perspective. A detailed examination of the structure of this concept, which will be referred to as the corporate memory system, will be included in the following sections of the study.

3 Typology of Corporate Memory

Corporate memory literature includes different categorizations of the term which are proposed by different researchers. Each categorization reflects a different viewpoint. Some of these are typologies based on direct observations from organizational setting while others are proposed based on the similarities to individual memory. In order to provide an understanding about the concept, the following categorizations are listed in this section.

3.1 *Semantic and Episodic Memory*

Semantic and episodic memory categorization is made in parallel to individual memory. Individual memory includes general facts and personal experiences (Croasdell et al. 1997). Semantic memory is the memory related with the knowledge of the facts about the world, while memory of the experience gained through events encountered over time called as episodic memory (Tulving and Craik 2000). On the organizational level, a parallel classification has been made. Accordingly, semantic memory refers to the memory of information about general rules and procedures in the organization (accounting records, standard operational procedures, etc.), while episodic memory refers to the memory of knowledge (specific strategic decisions, stories, etc.) that includes time and context regarding specific events related to the organization (El Sawy et al. 1986).

3.2 *Declarative and Procedural Memory*

Another classification related to organizational memory is the declarative and procedural memory. Declarative memory which is the combination of semantic and episodic memory carries information that is accessible. Procedural memory (also called skill-based memory), on the other hand, refers to the memory that carries the knowledge of how things are done (know-how), which can only be understood during practice (Stein and Zwass 1995).

3.3 *Public Access and Restricted Access Memory*

El Sawy et al., In their study in 1986, stated that not all of the organizational memory is accessible to all employees, and they made a different classification as publicly accessible and restricted access memory. By combining this classification with semantic and episodic memory classification, the researchers, who created a matrix with four different cells, suggested that different ways could be followed to manage the type of memory in each cell. Based on this matrix, accounting records of a company can be regarded as public access and semantic memory item, while stories or annual reports of the company are regarded as items of public access and episodic memory. When it comes to restricted access memory, SOPs (standard operational procedures) can be regarded as semantic and restricted access memory items, while details for strategic decision is an example of episodic and restricted access memory item (El Sawy et al. 1986).

3.4 Concrete and Abstract Memory

This memory classification has been made by Sandoe et al. (Stein and Zwass 1995). Concrete memory contains information on specific situations. Examples include the history of the organization, the state of its tangible assets or information related to customers. On the other hand, soft memory, such as technological or scientific information about the industry, is related with the knowledge of concepts and relationships without knowledge come from an experienced situation. For example, knowledge about Kant's understanding of ethical decision making can be considered as soft memory (Stein 1989).

3.5 Descriptive and Prescriptive Memory

This classification also appears as a classification based on the structure of the content within memory and is related with the normative orientation of knowledge. Descriptive memory refers only to the information that defines a situation, concept or relationship, in other words, the memory whose main purpose is to inform, while prescriptive memory refers to the memory related to the information that imposes certain or general rules for the behavior of the members of the organization (Stein 1989).

3.6 Communal and Idiosyncratic Memory

This organizational memory classification is concerned with whose perspective memory represents. Accordingly, organizational memory can be shared or it can be personal. Communal memory refers to the memory which consists of information obtained from groups that make decisions that affect organizations. The aim is to reach a consensus in views. Personal memory contains information provided separately from each decision maker. In this case, unlike the common memory, there are different, perhaps conflicting views stored and represented as alternative approaches for future relevant decision situations (Sowunmi et al. 1996).

3.7 Formal and Informal Memory

In this organizational memory categorization proposed by Harvey, Smith, and Lund, formal memory refers to the information contained in organization's records and systems. On the other hand, informal memory refers to the information organizational members get from their colleagues by using their personal relationships when it is needed. In this way, which is also called as mouth to mouth, organizational members utilizes from the memory of people they think have the necessary knowledge (Harvey et al. 1996).

3.8 Individual, Cultural and Prosthetic Memory

In this classification, it is stated that organizational memory is the combination of these three memories. Accordingly, individual memory refers to the memory of the information acquired from past experiences, cultural memory refers to the social elements that make up the collective identity, and prosthetic memory refers to the information contained in technology and not coming from the individual's own life experience. Individual memory can be affected by social processes from time to time and details are likely to be forgotten. It includes cultural memory, myths, stories, norms and rules. Prosthetic memory is carried by machines, tools, information and communication technologies (Martin Corbett 1997).

3.9 Memory Classification Based on Spatial and Temporal Integration

This classification proposed by Stein and Zwass is based on the spatial and temporal integration of organizational memory. Integration here is about the sharing of organizational knowledge over space (here, organizational space) or time. Spatial integration does not mean the gathering of organizational memory in one place; It refers to the interconnectedness of the memory scattered throughout the organization.. Temporal integration is about the transfer of knowledge between past, present and future. There are four types of organizational memory according to the levels of these two dimensions. These are classified as connected-amnesic (high spatial integration, low temporal integration), fragmented-amnesic (low spatial integration, low temporal integration), connected-retentive (high spatial integration, high temporal integration) and fragmented-retentive (low spatial integration, high temporal integration) organizational memory, and it is stated that the ideal memory is connected-retentive organizational memory. In addition to connecting the memory, which can be in different places within the organization, it should also be kept within the organization in a way that can be brought from the past in order to be used today and in the future. Otherwise, this organizational knowledge will be doomed to be forgotten (Stein and Zwass 1995).

3.10 Physical and Cognitive Memory

Physical and cognitive organizational memory classification is about whether information is stored in artificial carriers or it is implicit information (Eun Park and Bunn 2003). While programs, databases, structures and procedures make up the physical memory, beliefs, judgments and intuitions make up the cognitive memory (Anand et al. 1998).

3.11 Management, Technology, Culture and Market Oriented Culture

In this classification, there are four types of organizational memory. Three of these types namely management-oriented, technology-oriented and culture-oriented memory are from the internal environment of organization, while the fourth one namely market-oriented memory is from the external environment of the organization. Management-oriented organizational memory refers to the memory for the knowledge that controls the operation of an organization, such as management style or organizational structure. Technology-oriented organizational memory is the memory that includes technology and related experiences used to support and improve organizations' normal operations. Culture-oriented organizational memory is defined as the mental wealth accumulated during the development of the organization, such as the history of the organization, shared values. Finally, marketing-oriented organizational memory refers to the memory containing information about the market such as supplier, agent, customer-related sales and purchasing information (Li et al. 2004).

3.12 Expectation and Procedural Memory

Different from the previous classifications, another classification made about organizational memory is expectation memory. Expectation memory, expressed as the precondition of procedural memory which is specific to the organization and expresses the implicit stock of information about how things are done, is defined as the explicit or implicit expectations of the organization members from each other and from the whole organization (Conklin 1997). Accordingly, individuals need to have positive expectations that they will be rewarded in some way, today or in the future, to contribute to procedural memory (Ebbers and Wijnberg 2009).

3.13 *Transactive and Systemic Memory*

It is seen that the classifications made on organizational memory are generally made according to the type of information in the memory. However, another distinction that should be included among the classifications is a classification made for the level of analysis. In essence, in this distinction, which can be described as the memory of a group in the organization and memory for the entire organization, group memory has been named in different ways such as project, team or transactive memory (Stein and Zwass 1995; Lehner et al. 1998). Of these, transactive memory has been put forward by Wegner et al. Transactive memory, which is defined as a shared system for the acquisition, storage and retrieval of information, consists of a group of employees within the organization developing a common understanding with communication activities (Wegner 1987). Transactive memory also includes the information of from whom to access the needed information as well as the knowledge possessed by the individuals that constitutes the group (Wegner et al. 1991; Wegner 1987).

Systemic memory, on the other hand, refers to a memory related with the whole organization. In their study in 1998, Anand et al.. expanded the transactive memory and suggested that there could be more than one transactive memory in organizations and that these interconnected memories would form the systemic memory. Accordingly, systemic memory and group memories (in other words, transactive memory) are the expressions of the same concept at different analysis levels (Anand et al. 1998).

4 Debates on Organizational Memory

In the literature, there are different opinions on the concept of organizational memory, the relationship of organizational memory with other concepts and its possible negative outcome. In this section, these opinions, which include different perspectives on organizational memory, are examined under the titles of the of organizational memory concept debate, debate on the relationship of organizational memory with organizational learning and knowledge management, and organizational memory and rigidity debate.

4.1 *Organizational Memory Concept Debate*

Many concepts related to organizations have been tried to be explained based on individual characteristics. This approach, also known as the attribution of human traits to other beings, is called anthropomorphism (http://www.tdk.gov.tr/?option=com_karsilik&view=karsilik&kategori1=abecesel&kelime2=A). For example, organizational behavior, organizational learning or organizational intelligence were explained with anthropomorphism. However, in real sense organizations cannot think, learn or remember (Argyris and Schön 1978).

Similarly, organizational memory has been tried to be explained based on human memory. Discussions about the existence of organizational memory have a very long history, and human memory has often been used as a model and metaphor rather than organizational memory (Lehner 2004). Although organizations do not have a real brain, they do have cognitive systems and they must have memory to the extent that they have cognitive systems (Hedberg 1981; Walsh and Ungson 1991). This situation takes organizational memory beyond just being a metaphor used to explain the concept.

The concept of organizational memory has been tried to be explained biologically, psychologically or sociologically in various ways. In the biological point of view, it has been attempted to establish a similarity between the parts and functions of the brain and organizational memory. The human brain consists of several parts, each of which stores different types of memory. Wellman, in his 2009 study, linked different parts of the human brain to organizational memory with a biological approach. According to Wellman, organizational memory is stored in four different places in an organization. These are; culture, former employees, archives and processes. Culture refers to the set of operating principles and behaviors that are known to almost everyone in the organization, but not written. It creates an instinctive reaction to stimulus within the organization and is often rationalized but not understood. In other words, individuals exhibit certain behavioral patterns within the organization, mostly without questioning, in line with the culture of that organization. In this respect, culture has been associated with the amygdala, which is related to emotional memory in the human brain. This part of the brain directs the physiological stress responses initiated by the sympathetic nervous system, such as increased heart rate and blood pressure, dilated pupils, for example in a threatening environment. For example, although the way things are done in the organization is not written, it is clear and individuals act in this direction. The hippocampus, the part of the brain that manages long-term memory, has been associated with former employees, expressed as part of organizational memory. These individuals are those who have been in that organization for a long time to have a vast amount of experience in the products, processes, environment and capabilities of the organizations. So it has been shown to correspond to long-term memory. The cortex, the part of the human brain that stores facts and data, has been associated with archives, the best known repository of organizational memory, and finally, the cerebellum and basal ganglia that govern habits and routines has been associated with processes that carry knowledge of work routines and procedures, methods and techniques (Wellman 2009).

By offering a comparison with its key elements, the biological approach described above provides an area of understanding of what organizational memory can be (Atkinson and Burstein 2011). On the other hand, the psychological approach tried to explain organizational memory with psychological models of individual memory and focused on the process of bringing back past information (Martin Corbett 1997). Psychological models locate memory in the brain and see technology as a metaphor for organizational memory (Martin Corbett 1997). However, positioning memory only in the human brain is insufficient to explain organizational memory (Lehner 2004). It would be wrong to see organizational memory as the sum of individuals' memories. Therefore, a sociological perspective is needed (Martin Corbett 1997). According to this approach, the organizational interpretation process is more than what happens by individuals. Individuals come and go, but organizations over time store information, behaviors, mind maps, norms and values (Hedberg 1981). The distinguishing feature of organizational level information activity is sharing (Daft and Weick 1984). In this respect, it would be a more correct approach to consider organizational memory from a social perspective.

While there are still unclear issues about how people think and remember, it is not surprising that there are different views on organizational memory (Baddeley 1998). As a result, since the sociological theory of memory has a history of more than a century and the developments in information processing, artificial intelligence and technology (Lehner 2004) have made the concept of organizational memory a more understandable and investigatable subject.

4.2 Debate on the Relationship of Organizational Memory with Organizational Learning and Knowledge Management

One of the debates about organizational memory is the relationship of the concept with organizational learning and knowledge management. While some researchers consider organizational memory as a meta concept that includes organizational learning and knowledge management (Lehner 2004), there are also those who accept that knowledge management includes organizational memory (Burstein et al. 1998; Hatami and Galliers 2005; Jasimuddin et al. 2009). Regarding the relationship between organizational learning and organizational memory, many researchers agree that these two concepts are intertwined concepts that feed each other (Hatami and Galliers 2005), while Hanvanich, Sivakumar, and Hult suggested that organizational memory and organizational learning are independent from each other (Hanvanich et al. 2006). Spender, on the other hand, has handled the concepts of organizational memory, organizational learning and knowledge management together, and he mentioned the necessity of considering these three concepts in relation to each other, like the relationship between voltage, current and resistance in Ohm's Law (Spender 1996).

Organizational memory and knowledge management are well known concepts from organizational science and learning theory. These concepts are seen as tools that offer a systematic application to the way the organization processes information (Lehner 2004). Organizational memory should support knowledge creation and organizational learning in an organization in the long run (Abecker et al. 1998). Organizational learning is not the sum of individuals' learning, but an organization learns through individuals (Argyris and Schön 1978). Therefore, it is basically a process in which the individual is involved. If what has been learned at the end of a problem solving process is not stored for future reuse, the learning will have to occur from the beginning when the problem is repeated. However, even if it is stored, it will not be possible to speak of a sufficiently effective learning unless what has been learned/experiences can be brought back when needed (Koornneef and Hale 2004). Therefore, just as it is impossible to talk about learning without memory, organizational memory is a prerequisite for organizational learning (Lehner 2004). Organizational memory is at the core of learning organizations and enables the sharing and reuse of individual, organizational knowledge and lessons learned (Abecker et al. 1998). While Walsh and Ungson argued that improving memory and especially updating it increased learning (Walsh and Ungson 1991), Argyris and Schön also stated that organizational memory is necessary for organizational learning (Stein and Zwass 1995). Because for organizational learning to take place, the discoveries, inventions and evaluations of learning agents within the organization must be embedded in organizational memory and retrieved when necessary (Argyris and Schön 1978).

The main function of organizational memory is to increase competitiveness by improving the way the organization manages information (Abecker et al. 1998). Therefore, researchers trying to understand knowledge management can take organizational memory models and terms as a starting point (Anand et al. 1998). Knowledge management is a comprehensive concept expressed as the management of implicit or explicit information within and outside the organization (Wegner et al. 1991). This concept, which is also called second generation knowledge management today, actually refers to the management of all information processes within the organization (McElroy 2003). When considered in this way, the importance of organizational memory in terms of knowledge management is obvious. As a matter of fact, Jennex and Olfman talked about the relationship between organizational memory, organizational learning and knowledge management in their 2002 study, and argued that these concepts support each other and also have an impact on organizational efficiency (Jennex 2005).

As a result, if knowledge management is considered as a meta-concept in the form of "management of information processes" from today's perspective instead of its traditional definition, organizational memory and organizational learning are also interacting with each other under the umbrella of knowledge management, just like all processes that must be carried out in interaction with each other for the effectiveness of the organization and they can be described as an information process managed in a way that supports each other.

4.3 *Organizational Memory and Rigidity Debate*

Another controversial issue regarding organizational memory is whether the information kept in memory creates a barrier for the organization. While there are those who advocate the usefulness of organizational memory, there are also researchers who completely ignore it or see it as an obstacle to change (Martin Corbett 1997). In fact, the basic assumption of researchers who argue that organizational memory is useful is that the information stored in memory is always valid (Croasdell et al. 1997). However, it is possible that the conditions will change in the time period between the moment the event that is the subject of experience and today, and the information memorized under these conditions will lose its importance for today's decision making process. If only a short time has passed between the present and the event, this difference may be small, but as the time passes, it will be difficult to establish the relationship between the past and present conditions. As the time passes, it will be difficult to perceive the meaning of the experience gained like that of those who have experienced the event (Ackerman 1994). However, it would also be wrong to assume that past events and experiences have little relevance to present-day decisions due to rapidly changing circumstances (Kransdorff 2006). The main thing here is to know how the past can best be harmonized with the future (Kransdorff 2006), what should be remembered or forgotten (Tuomi 1996). In other words, the basic condition for the effectiveness of organizational memory is that the information in the memory can maintain its validity over time. The way to achieve this is through updating (Wegner et al. 1991). A memory that is not updated will carry information that has lost its validity and value for today's conditions, and may lead to decisions based on rigidity and false information, as stated by researchers who claim that organizational memory prevents change (Stein and Zwass 1995; Argote 1999). Therefore, an organizational memory must be adapted to emerging requirements, constantly updated and improved so that it does not turn into a disadvantage for the organization (Abecker et al. 1998). While this was difficult to achieve in the past (Argote 1999), with the advancements in information technologies and its widespread availability, information systems have become a vital part of organizational memory (Stein and Zwass 1995), which has turned it into an advantage for the organization by allowing to update the disadvantageous aspect of organizational memory.

5 The Importance of Corporate Memory and Organizational Results

Corporate/Organizational memory is not only a reason for being for all organizations, from private businesses to public enterprises, but also a key to the survival of the business (Kransdorff 2006). As a matter of fact, the ability of an organization to collect, transfer or use the information obtained from its experiences can provide a stronger and faster learning (Wexler 2002), successful new products/services (Moorman and Miner 1997), more effective decision-making (Croasdell 2000), and consequently a higher organizational performance (Li et al. 2004).

Although organizational memory is so important for organizations, studies on this subject have been limited (Pratt 2006). The subject has started to gain the importance it deserves with the realization that learning from experience is a key element for competitiveness (Stein and Zwass 1995). When the empirical studies on corporate/organizational memory are examined, it is seen that these studies are generally critical issues for organizations such as organizational performance, innovation, competitive advantage.

One of these studies was carried out by Johnson in 1993. With this research using experimental design, the issue of whether the knowledge of past decision-making behaviors would be useful in current decision processes was investigated. With this research conducted on students at the University of Arizona, it was concluded that the use of information about past decisions in current decision processes is useful (Johnson 1993).

A study conducted by Moorman and Miner in 1997, in which the impact of organizational memory on new product performance and creativity was investigated, was carried out with the data obtained from the product development project of 92 companies that are in the first 200 of the 1992 Advertising Age list. In this study, organizational memory is defined as changing collective beliefs, behavioral routines, and physical artifacts in terms of content, level, distribution and accessibility. Organizational memory diffusion refers to the level of sharing of organizational memory among individuals. In the study, short-term new product performance refers to the sales and profitability of the product in the first year of its launch, and new product creativity means that the product has a special and productive capacity. As a result of the research, it was concluded that organizational memory level increases the short-term financial performance of new products and organizational memory spread increases both new product performance and creativity (Moorman and Miner 1997).

Paul et al. studied the effect of organizational memory on cognitive-conflict decision-making with their study in 2002. In this study conducted by using the control and experimental groups, the relationship between the knowledge of past decisions and decision making in the evaluation of applications to the MBA program was examined. As a result of the research, it was concluded that organizational memory affects the speed of the decision positively (Paul et al. 2004).

Another study on organizational memory carried out by Li et al. This study, which investigates the effect of organizational memory on organizational performance in manufacturing enterprises, was conducted on the basis of 2001 data from the International Manufacturing Strategy Survey (IMSS). The International Manufacturing Strategy Research is a worldwide research project and covers 600 businesses from 20 countries. This project that was first implemented in 1993 by the London School of Business and Chalmers University of Technology was designed to identify and explore the strategies and practices of countries around the world. In the study, organizational memory was generally approached from the perspective of the use of information and examined in four dimensions: technical, managerial, cultural, marketing organizational memory. Technical organizational memory is professional knowledge-based memory that supports and improves the normal operations of the organization, including technology and related experiences. Managerial organizational memory refers to the information that controls the operation of an organization. Cultural organizational memory has been defined as the mental wealth accumulated during the development of the organization. Marketing organizational memory, on the other hand, refers to the information about suppliers, intermediaries and customers in the field of sales and purchasing. Organizational performance is similarly examined as technical, managerial, cultural and marketing organizational performance. Technical organizational performance refers to performance in the production process, managerial organizational performance refers to the performance in elements related to the management of the enterprise, cultural organizational performance refers to the relationship between employees and the enterprise, and the connections between members in an organization. Cultural organizational memory and cultural organizational performance were not included in the analysis due to lack of measurement. As a result of the research, it was concluded that organizational memory on the basis of general variables has a significant positive effect on organizational performance. In the analysis made on the basis of dimensions, it was seen that technical and managerial organizational memory had a positive and significant effect on both managerial and marketing organizational memory, while marketing organizational memory had a positive and significant effect only on marketing organizational performance (Li et al. 2004).

In the study conducted by Hanvanich, Sivakumar and Hult in 2006, the effect of learning orientation and organizational memory on organizational performance and innovation and the moderator role of environmental change in this effect were investigated. In the study, learning orientation is defined as the level of organization's adoption of organizational learning as a set of basic values necessary to survive. Organizational memory is expressed as the stored information of the enterprise. While environmental turbulence is expressed as the amount of change of key environmental variables and the unpredictability of the future level of these variables, it is included in the study as technological turbulence and market turbulence. Technological turbulence is determined as the level of change in product and process technologies in the industry with which the company is associated, and market turbulence is the rate of change in all of the customers and their preferences. Among organizational results, innovation is defined as the ability of businesses to research novel ideas, accept innovation and support idea development. In this study of 200 supply management specialists working in production enterprises, one person from each business, as a result of the research, it has been observed that learning orientation has a positive effect on both organizational performance and innovativeness, and this effect is stronger in environments with high technological and market turbulence than in environments with low level turbulence. On the other hand, it was seen that organizational memory has a positive effect on organizational performance and innovativeness, but here, the result is opposite of the result related with learning orientation. Accordingly, it has been observed that the effect of organizational learning on both organizational performance and innovation is weaker in environments with high technological and market turbulence than in environments where it is low (Hanvanich et al. 2006).

The 2010 study of Camison, Boronat and Villar examined the mediating role of knowledge-based skills in the effect of technological strategic alliances on organizational performance. In this study, the stock of knowledge is described as organizational memory and expressed as a knowledge-based skill. The stock of information has been defined as the valuable knowledge of the business and organizational performance, on the other hand, was considered as economic performance. As a result of this research conducted with the CEOs of 401 businesses randomly selected from the Spanish industrial enterprises registered in the ARDAN database, it was concluded that the information stock has a positive and significant effect on organizational performance (Camison et al. 2010).

Another study on the effect of organizational memory on organizational performance was conducted by Lee et al. In this research conducted on 610 participants from 22 health centers in Taiwan, it was concluded that individual and organizational memory has a positive effect on hospital health care performance (Lee et al. 2011).

Lai et al. investigated the effect of organizational memory on employee service performance. This study, which was conducted in 2011, was carried out on 256 healthcare workers and 34 supervisors working in 34 hospitals operating in Taiwan. As a result of the research, it was seen that organizational memory has a positive effect on employee service performance (Lai et al. 2011).

Another study on organizational memory was conducted by Dunham and Burt. This study, which was conducted in 2011 to determine the effect of organizational memory on psychological strengthening, was conducted on 134 employees representing 6 enterprises. As a result of the research, it was concluded that organizational memory has a positive effect on psychological empowerment (Dunham and Burt 2011).

Camison and Villar's study was conducted on the impact of organizational memory on non-technical innovation (organizational and marketing innovation). In this study conducted on 159 industrial enterprises in the database of SABI (Sistema de Análisis de Balances Ibéricos) in Spain, as a result of the research it was seen that organizational memory has a positive effect on both types of innovation (Camison and Villar 2011).

As can be seen from the studies on the subject, organizational memory has positive effects on issues such as empowerment, innovation, organizational performance required for businesses to maintain their competitive advantage and to survive.

6 Basic Models of Corporate Memory

There are many different perspectives and classifications in the literature regarding corporate/organizational memory. However, there are few studies that deal with organizational memory from a holistic perspective. For this reason, in this section, basic models proposed for organizational memory will be included. The first detailed examination of the structure of organizational memory was made by Walsh and Ungson in 1991. Following the work of Walsh and Ungson, Stein and Zwass raised the issue of the support of information technologies on organizational memory in 1995 and proposed a model for the organizational memory system. Finally, in 1998, Anand, Manz, and Glick analyzed organizational memory from the perspective of the unit of analysis, expanding the group memory and included an organizational memory belonging to the whole organization in their model.

6.1 Walsh and Ungson's Organizational Memory Model

Walsh and Ungson defined the concept of organizational memory with their work in 1991 and talked about the processes of organizational memory. In this study, organizational memory is expressed as "stored information related to the history of the organization and that can be used in current decisions" (Walsh and Ungson 1991).

Walsh and Ungson approached the organizational memory from the perspective of storing, as can be seen from the definition in the model they proposed (Spender 1996). According to this model, organizational memory consists of the processes of acquisition, storage and retrieval of information, and the memory in enterprises exhibits a structure in which information is stored in more than one place rather than a center (Van Rensburg 2011). These repositories, also named as storage bins, are described as individuals, culture, transformations, structures and ecology (working environment), and although external archives are not in the business, they can be a source of information for the history of the business, so they are included in memory carriers (Walsh and Ungson 1991).

The acquisition from organizational memory processes is closely related to information about a specific decision stimulus (this may be a problem or a need for change) or the consequences of an applied decision. Information about a stimulus that initiates the decision-making process is usually acquired by individuals and the organization provides an organization response to this stimulus (Wellman 2009). As a result, the interpretations of organizational decisions and the results of these decisions form the memory of an organization (Walsh and Ungson 1991).

The acquired knowledge is stored in different places within the organization. One of the places where knowledge is carried in the organization is ***individuals' own memories***. Individuals have their own knowledge of events that occur within or related to the organization. Individuals store their organization's memory/knowledge in line with their own capacity to remember and create experiences (Bannon and Kuutti 1996) and their cognitive orientation that they use to facilitate information processing. Another carrier where organizational memory is stored is expressed as ***culture***. For organizational culture, Schein's definition in a study in 1984 was adopted and it was expressed as the way of perception, thinking, feeling learned about problems and transferred to the members of the organization (Walsh and Ungson 1991). Walsh and Ungson drew attention to the words 'learned' and 'transferred' in this definition (Wellman 2009) and characterized organizational culture as a carrier of organizational memory in terms of containing past experiences that may be useful in dealing with future situations (Walsh and Ungson 1991). Accordingly, many cultural elements, including language, symbols, and stories, contribute to collective knowledge. Especially in terms of decision-making, past experiences manifest themselves in organizational culture (Van Rensburg 2011). Another organizational memory carrier is ***transformations***. Whether this is the transformation of the input of a physical product into an output, or any process taking place within the organization (such as the recruitment process), information is involved in many transformations that take place within the organization (Walsh and Ungson 1991). In fact, this refers to the information that drives the transformation processes taking place (e.g. procedures and rules) (Bannon and Kuutti 1996). Accordingly, for example, standard operational procedures for past transformations within the organization form a guiding scheme and become a part of organizational memory, which can be useful in future processes. As an organizational memory carrier, organizational structure should be considered in light of its possible consequences for individual role behavior and its connection with the environment. As a sociological concept, ***roles*** are related to the classification of certain positions in a society and are based on social expectations. Similarly, there are various roles within the organization and these roles are closely related to the organizational structure. Therefore, the concept of role provides a link between individual and organizational memory and is a carrier of memory within the organizational structure (Walsh and Ungson 1991; Bannon and Kuutti 1996). Information is encoded in roles and drives behavior (Van Rensburg 2011). ***Ecology (Working Environment)***, or in other words, the physical arrangement of the business is an important source of information about the business. Physical arrangement mostly reflects the status hierarchy within the organization and helps shape behavior (Walsh and Ungson 1991). For example, properties such as owned office area are an indicator of the status of employees, and the resulting roles and behavior patterns form organizational memory (Van Rensburg 2011). Therefore, ecology stores information about an organization and its members. Finally, although ***external archives*** are not a part of organizational memory, they are sources from which past information about the organization can be retrieved. This could be a former employee, information acquired by competitors or data obtained by a valuation agency (Walsh and Ungson 1991). Information retrieval from organizational memory processes is also expressed as using the information contained in organizational memory carriers while responding to a certain decision stimulus (Walsh and Ungson 1991).

According to this model, the acquisition of information is performed by individuals and the memory is stored in five carriers within the organization. Carriers that make up organizational memory, including external archives, are used to support the decision environment. At this point, retrieving information, another process of organizational memory, comes into play. In this direction, the information stored in the organizational memory is used in current decision-making processes when needed.

The model fell short on the role and importance of information technologies for organizational memory. In this respect (Stein and Zwass 1995), it is thought that for today's complex organizational structures, this situation may cause difficulties in meeting organizational needs. As mentioned in the debates on organizational memory, updating the organizational memory with the help of information analysis it offers for the organizational memory information system. In the model, which is limited in terms of the social aspect of organizational memory, handling the issue from a system perspective constitutes an important step in the structure of organizational memory.

6.3 *The Organizational Memory Model of Anand, Manz, and Glick*

Another model on organizational memory was developed by Anand, Manz and Glick in 1998. In this proposed model, based on group memory (transactive memory), an organizational memory for the whole organization is mentioned. In this study, which draws attention to the level of analysis regarding organizational memory, individual memory, group memory and systemic (for the whole organization) memory are included (Anand et al. 1998). In this model, which also stresses the separation of internal and external memory, it is stated that similarly interconnected group memories form organizational (systemic) memory, just as the sharing and communication of individuals are important for the formation of transactive memory.

According to Anand, Manz, and Glick, systemic and group memories actually represent different levels of analysis of the same concept. However, while the systemic memory of an organization contains large amounts of information, some group memories may be incomplete because they cannot access the information contained in the systemic memory due to poorly developed directories and/or communication barriers encountered by group members. Or, some groups (e.g. a product development team) may have very knowledgeable members, while the organization may have relatively less knowledge (Anand et al. 1998). Important here are the links between different levels of memory.

Anand, Manz, and Glick also touched upon the impact of organizational memory on decision making, as in other models. Accordingly, individuals cannot have all the information necessary for the decision-making process and they have to obtain this information. For this, they need to make use of organizational memory. Individual knowledge of organization members is also part of organizational memory and should be made accessible with locators. In this way, the decision maker will be able to access the source of expert knowledge that he does not have individually and ensure that this information is used in the decision-making process (Anand et al. 1998).

In addition to the common aspect of organizational memory in decision-making support, unlike previous models, the issue of analysis level was handled with this model and contributed to the issue of organizational memory. However, while the social aspect of organizational memory was not addressed in other models, the subject of communication was emphasized in this model, but issues such as interpersonal trust or reliability of information were not addressed.

7 Organizational Memory as a Socio-Technical System

Individual memory is structured through collectively held ideas and experiences shared with others. Most social experiences and interactions are through technology. Therefore, memory is not entirely personal. It is also social and partly technological (Martin Corbett 1997). Like individual memory, organizational memory is not a concept that can be considered independently of the social environment. If a collective memory is mentioned, the individual, the social environment that provides the interaction of individuals and technology support should be considered together. As a matter of fact, in a definition related to organizational memory, it was stated that the concept includes the technical, functional and social aspects of the worker, the workplace (Croasdell et al. 1997).

Organizational memory has been studied socially by some researchers (http://www.tdk.gov.tr/?option=com_karsilik&view=karsilik&kategori1=abecesel&kelime2=A; Wellman 2009) and technically addressed by some researchers (Burstein et al. 1998; Attipoe 1999). In fact, organizational memory is a function of social mechanisms and technologies (Conklin 1997), and it would be more appropriate to handle a concept that integrates technical artifacts with social processes (Tuomi 2000) by a system viewpoint.

In its most general definition, the system refers to a whole that consists of interrelated subsystems and is also connected with the upper system. Ludwig von Bertalanffy's general systems theory is also the basis for the development of the organizational memory system (Pratt 2006). Because, the organizational memory system consists of components that ensure the functioning of the system such as the individual, social environment, and technology and that interact with each other. And it is also in a position which is connected with other systems inside and outside of the organization.

Miller suggested that organizational memory consists of human and non-human components and mentioned written documents, books, photographs, voice recordings, filing systems, and computers as examples of non-human components. He also mentioned the record keeping and computer filing departments as examples of the human component (Stein 1989). Huber, on the other hand, described human as an organizational memory component and drew attention to the information technology component of organizational memory against the difficulties that may arise due to reasons such as labor turnover and reluctance to share information (Huber 1990). Ackerman and Halverson also dealt with the organizational memory as a whole with the human element and technology (Ackerman and Halverson 2000).

Stein suggested in his 1989 study that organizational memory is a socio-technical system (Stein 1989). Many researchers also support this view (Harvey et al. 1996; Jasimuddin et al. 2009; Lynne Markus 2001; Stoyko 2009). Organizational memory means more than the sum of individual memories beyond the individual memories of organization members (Wexler 2002). It is shaped by social interaction and has a collective nature. It is social in this respect. However, organizational memory has always been supported by a number of technical elements from its simplest form to today's advanced information technologies. This can be a written source, document, filing system or an electronic database. In this respect, it is a technical concept. Therefore, in this chapter, organizational memory is considered as a socio-technical system.

8 A New Model for Corporate Memory System

When the models suggested before regarding organizational memory are examined, it is seen that these models differ in certain aspects. For this reason, a new organizational memory model to be created based on these differences, which can be considered important in terms of the structure and functioning of the organizational memory system, is important.

Organizational memory does not have a single and centralized structure. The sources of organizational memory are varied. Sometimes the memory used is individual and private, sometimes group and public. However, all scattered memory carriers must be used in unity for a common purpose (Ackerman and Halverson 2000). If the memory is stored in different places within the organization, but there is no connection between these memory carriers, the organizational memory contributes very little to the organization. If the information stored in the organizational memory is easily lost, it is not possible to talk about a useful organizational memory. For this reason, organizational memory should be both linked and conservative (Croasdell et al. 1997). This is possible with the integrity of the components that make up the organizational memory system.

Since a part of the knowledge required for the organization is in the minds of individuals and individuals are the most effective means of obtaining and storing implicit knowledge, one of the components of the corporate memory system is the ***human factor***. However, being completely dependent on the individual in corporate memory can also have negative consequences for the organization. The most important of these is that individuals leave the business with the knowledge they have. For this reason, it is necessary to consider the ***supportive environment***, which is another component of organizational memory, in order to retain the human factor, which is a valuable asset for businesses as it carries experience that cannot be turned into open knowledge, and to remove information from individuality as much as possible and give it a collective character. However, in the corporate memory system, human and technology are complementary to each other. With ***information technologies***, it will be possible to make the information contained in the corporate memory usable in the processes of the organization in the business environment. In addition to the human factor, supportive environment and information technologies support, another component of the corporate memory system is ***written-audio-visual (WAV) resources***. Written sources such as important information about an organization, procedures, and reports can be included in various audio or video recordings. Accessibility of these resources is also important for the corporate memory system. All of these components together

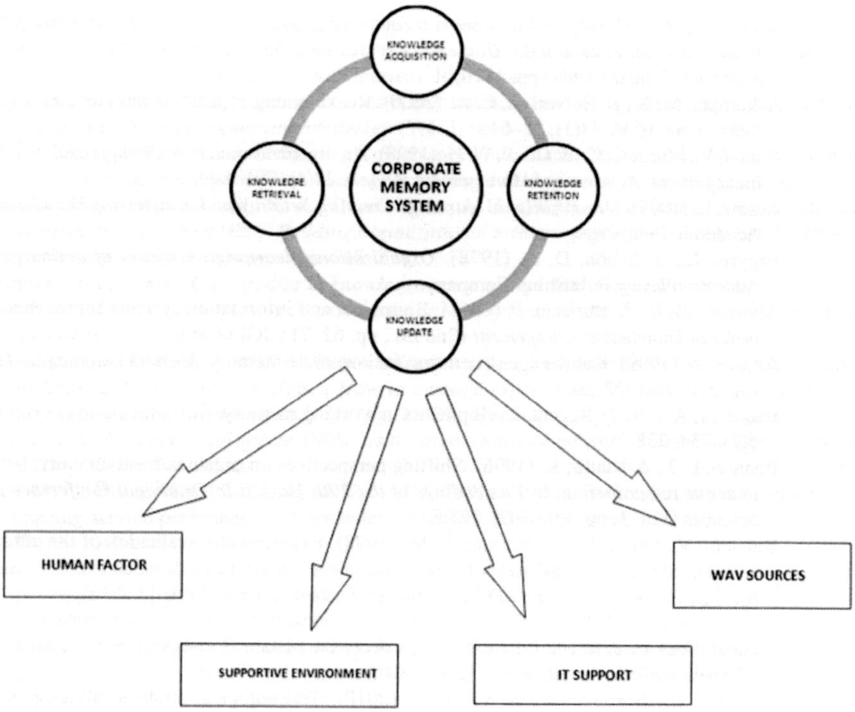

Fig. 1 Sen's corporate memory system model

form the corporate memory system and ensure the realization of corporate memory processes (Fig. 1).

The knowledge acquisition from corporate memory processes is about taking all kinds of information that may be needed in business processes from inside or outside the organization to the corporate memory system. ***Knowledge storage***, as the name suggests, refers to the storage of information in the corporate memory system. ***Knowledge update*** refers to updating and organizing the information in the corporate memory system against the risk of losing its validity over time, while ***knowledge retrieval*** means retrieving the information in the corporate memory system when needed.

References

Abecker, A., Bernardi, A., Hinkelmann, K., Kühn, O., & Sintek, M. (1998). Toward a technology for organizational memories. *IEEE Intelligent Systems*, May/June 1998, 40–48.

1 Introduction

An organization's ability to learn can be a factor of competitive advantage (Odor 2018). For this, it is important that the organization is available to understand the new challenges of the contexts in which it operates and the needs for change that the current dynamics in the organizational world imply. This learning dimension of an organization is closely linked to the organizational culture, leadership styles, values and principles that it advocates and implements in practice and the need for continuous improvement. There are organizations that, when they reach a certain type of performance level, lose or become distracted by the need to continue learning and end up failing and even disappearing (Appelbaum and Goransson 1997; Caldwell 2012; Odor 2018; Ribeiro and Gomes 2016, 2017; Senge 2003). The same is true for professionals in any field who fail to perceive continuous learning as a guarantee of survival in a world in constant change (Handy 1992).

The development and construction of success profiles aims to hold the organization, as well as the professionals that integrate it, accountable for understanding its role in the process of continuous learning and availability to learn. The contextualization of this concept in HRM functions has to do with the fact that this is the area of management in organizations where continuous learning has a formal and organized existence, and is the area from which functional profiles are expected to be defined.

An HRM system includes different functions and all of them add value in different ways to the organization. HRM practices such as job description and analysis, recruitment and selection, training and development, performance evaluation, career management, among others, need and must be integrated into business strategies, since failure at the HRM level can lead to failure of a business plan, even if very well structured (Kaufman 2001; Markoulli et al. 2017; Mitsakis 2014). In addition, it is important to be aware that in the last three decades HRM practices have undergone constant changes and, therefore, it is also important that a human resources (HR) professional is constantly updated (Markoulli et al. 2017; Ulrich and Dulebohn 2015).

Research on knowledge, skills and behaviors necessary for HR professionals to be efficient and successfully perceived in terms of their performance, in order to make other professionals effective, began in the late 1970s (Cohen 2015; Sanders et al. 2014; Tornow 1984). Previously, research has focused on analyzing and describing functions, that is, obtaining information about the functions, evaluating that information and using it to determine the skills and competencies needed for the respective function. The analysis, description and qualification of functions is still one of the central and structural processes of HRM, aiming to support recruitment and selection processes, training systems and processes, performance evaluation policies and systems, career management processes and even definition of remuneration, benefits and incentives systems (Armstrong and Taylor 2014; Cohen 2015; Jaiswal 2018; Melo and Machado 2015).

In the 1980s, the first competencies models introduced by Ulrich began to emerge. This author started a longitudinal study on the skills of HR professionals that is still used today. A competency model represents a strategy that defines both the competencies of the company as well as the people who are part of it so that, together, they can achieve results (Munck et al. 2011). Thus, the creation of competency models has also generated the term *competency management*. Competency management maps the organization's competency needs, identifies competency gaps and fills these gaps from sourcing, training and development related activities, such as coaching and other corporate education activities (Barbosa et al. 2015; Graen et al. 2020).

Therefore, the concepts of *analysis and job description* and *competence management*, despite being different, are interconnected. The analysis of the function, as well as its description, is fundamental, since until the behavioral requirements of a function are not identified, the selection processes, training programs and performance evaluation systems cannot be developed in a sustainable way (Hassan 2016; Landy and Vasey 1991; Rego et al. 2015). The analysis of functions allows to check information that results in decision making in relation to HR planning, allowing to obtain information about individual competence: knowledge, skills and attitudes (Munck et al. 2011). In the case of recruitment and selection processes, it is based on the analysis of functions that job advertisements are built and guidance is provided to employees, informing them of the positions they will occupy (Rego et al. 2015). Job analysis focuses on describing and measuring work tasks while competency models create a common thread that influences employees' performance in their daily lives, according to the company's strategy. An analysis of functions aligned with a competency model will generate better results in human resources applications and practices (Armstrong and Taylor 2014; Sanchez and Levine 2009; Rego et al. 2015).

The link to career planning and management is also evident as the development of careers can be built according to the analysis of functions of an organization, since this analysis defines professional groups and the articulation between the functions and the development processes and organizational changes. In fact, these three concepts—analysis and job description; competencies management; career planning and management—influence each other (Armstrong and Taylor 2014; Rego et al. 2015).

The introduction of profiles of success, a new process in the HRM area, allows the company to combine analysis and job description, competence management and career planning and management, thus generating guidelines for HRM processes based on the study of critical experiences and knowledge. The creation of a profile of success turns out to be more detailed as it combines the areas that need to be developed to achieve success with the identification of the logical and mental capacities that are important. In the critical *success profiles*, it is also possible to identify the *talent from*, the *talent pool* and the skills necessary for the function, hence its connection with career management and competencies management. This development will contribute to facilitate HRM practices, more specifically the recruitment and selection process, as well as competencies management and career management, since these profiles include *traits*, *skills*, *talent from* and *talent pool*.

This chapter aims to provide an analysis of the relevant HRM practices that can contribute to build a success profile centered and contextualized in HRM professionals. Through a literature review of key theoretical concepts—analysis and job description, competency management, career and talent management—the study provides a better understanding that the construction of success profiles result from the use and articulation of HRM principles and practices and their interconnection. Therefore, the study aims to answer to the research question: How can the elaboration of a success profile integrate the analysis and job description in order to predict a better performance of the employee and contribute to the continuous improvement of other HRM practices?

This chapter is structure in eight sections, as follows. After the instruction, the context of organization and HRM is described. This section is followed by the description of Ulrich's HR competency model. The fourth section provides a narrative of the analysis and job description. Next is analyzed the main aspects of competence management, which is followed by an analysis of career and talent management. The concepts analyzed in the previous sections are integrated to provide a better understanding of success profiles, in the seventh section. The chapter ends with the conclusion.

2 The Context of Organization and HRM

The creation of an organization is one of the ways that human beings use to coordinate actions and/or acquire something they aspire to or value. Organizations can be an answer and a means to satisfy some human need. In addition, it can be defined as a unit that is constituted by human, material and financial means and that is conditioned by environmental variables (Lisboa et al. 2004). The organization creates value from inputs of material resources, such as input from raw materials; from the environment, such as shareholders, managers, customers, suppliers, trade unions, government; from transformation processes, such as the implementation of new technologies; and, still, from outflows of resources, such as salaries or dividends (Jones 2010; Ribeiro and Gomes 2016).

Organizations work according to some dimensions, such as the structure, design and organizational culture. The structure is a "formal system of tasks and relationships of authority that controls how people cooperate and use resources to achieve organizational goals" (Jones 2010, p. 6). In addition, it is usually associated with a stable operating framework for the organization (Lisboa et al. 2004). Organizational culture, on the other hand, involves the "set of shared values and norms that control the interactions of members of the organization with each other and with people outside it" (Jones 2010, p. 6). It can also be considered as an orderly and emergent system of meanings and symbols shaping the way members of an organization interpret their experience and act on an ongoing basis (Cunha et al. 2016, p. 622). Finally, organizational design is the process by which managers choose and manage various dimensions and components of culture and organizational structure, so that the organization can control the activities necessary to achieve its objectives. (Jones 2010). Organizational culture is the main differential of successful companies and even though it is not the only factor that affects the success of an organization, the development of a corporate culture surpasses factors such as corporate strategy, market presence and technological advantage (Arditi et al. 2017). Professionals now recognize that organizational culture has a strong impact on long-term performance (Arditi et al. 2017). Thus, the organizational culture requires a structure, which is related to control and conditions the options and decisions of managers, while the management process is based on planning, organization, evaluation, control and leadership (Barney 1986; Lisboa et al. 2004; Ribeiro and Gomes 2016). Therefore, the objectives of a structure are as follows: to identify formal reporting relationships, to group individuals by departments and to determine the relationships between those departments, through coordination and integration processes based on efficient and adequate communication systems (Cunha et al. 2016).

In addition, an organization is usually defined by context (age and size, technology, environment and power), structure (control mechanisms and design parameters), mentioned above, and processes (communication processes, leadership, and decision-making). The structure varies according to the complexity, formalization and centralization and usually follows the organizational strategy, and its determinants are the dimension, existing technology, environment and strategy (Cunha et al. 2018). Thus, organizational structures can be designed in several ways: functional, divisional, matrix and network (Jones 2010; Lisboa et al. 2004).

In addition, in the seventies, some authors, such as Child (1972), report that the design of the organizational structure has a limited effect on the levels of performance achieved. They add that the professionals responsible for delineating the context must know how to deal with change, in order to retain the preferential structure without compromising performance. Therefore, the structure's contribution to the organization's overall performance levels is indisputable, since an effective allocation of tasks and functions, such as the existence of a coordination system, ensures work with the objective of maximizing efficiency (Lisboa et al. 2004). Furthermore, since the organizational structure always follows the company's strategy, Parmenter (2010) refers that strategy is a way that the organization has to fulfill its goals.

In a competitive environment, the strategy will define the type of competition and will determine what the best resources to achieve the desired results and the desired performance are (Parmenter 2010). It is also important to note that, currently, the performance of organizations is dependent "on their flexibility and adaptability, in order to ensure an effective adaptation to a context that changes at an ever faster pace and with an intensity that never stops" (Lisboa et al. 2004, p. 224). In this context of constant changes, HRM turns out to be a very important agent (Handy 1992). Huselid, Jackson and Schuler (1997) state that it is important to define a set of practices that allow the organization to achieve strategic objectives. For example, if the organization decides on a strategy focused on low cost, HRM practices will be directed as follows: performance evaluation with a focus on control over short periods of time and specialized functions. If the organization decides on an innovation strategy, it will focus on an evaluation over long periods of time, with jobs determined in a less specialized way (Kapoutsis et al. 2019).

Initially, the HRM was more connected to aspects of a more bureaucratic-administrative character and without much connection to the business (1900 to 1960).

Between 1960 and 1980 the *personnel administration* gave way to the *personnel function*, which became part of the companies' organizational chart and began to develop not only administrative and legalistic issues within the scope of control, but also incorporated some of the practices that have become usual in HRM. After 1980, the HRM designation came to be used and gradually came to have a more articulated performance with the business and the organization's strategy (Ribeiro 2014). Over the years, this area of management has become more integrated and also more strategic to the point that, in some organizational realities, it takes on an effectively leading role. However, there is still a considerable gap between what is considered, in terms of discursive rhetoric, to be the role of HRM and then what is seen in many organizational realities regardless of size, culture, or geography (Cabral-Cardoso 1999, 2004; Jaiswal 2018; Legge 1995; Ribeiro and Gomes 2016).

In general, HRM today is related to all management decisions that affect the relationship between the organization and employees. Thus, the HRM system is closely linked to the organizational structure and the company's mission, strategy, political, cultural and economic environment. In addition, the main objectives of the HRM are: to support the organization to achieve goals, to efficiently employ the workforce, to provide the company with well-trained and motivated employees and to amplify maximum satisfaction by developing quality of life at work (Ribeiro and Gomes 2016). In this way, HRM will enhance strategic HR management that will allow the integration of essential objectives, policies and procedures, forming a coherent unit (Ivancevich 2008; Kapoutsis et al. 2019; Ribeiro and Gomes 2016).

In conclusion, strategic HRM is at the intersection of four elements: organization, people, HR systems, and organizational culture; with an interaction between strategy, objectives, policies and standards, the HRM becomes a fundamental element for the success of any organization (Caldwell 2003; Ivancevich 2008; Kapoutsis et al. 2019; Ribeiro and Gomes 2016). This leads to the relevance in starting to think about the development of job profiles for the functional and hierarchical responsibilities of this area of management.

3 Ulrich's HR Competency Model

As previously mentioned, HRM is closely linked to the organization's management, which is why competence models have started to appear, with the aim of studying what competencies are needed by HR professionals, in order to allow the sustainable growth of an organization.

Therefore, an HR professional who wishes to constantly improve his/her role needs to clearly understand the evolution of contexts, the evolution and change of organizational systems, the change in the conception of work and how to adjust, adapt, develop and update/recycle. (Kapoutsis et al. 2019; Markoulli et al. 2017; Ribeiro and Gomes 2016). Since 1987, different taxonomies have been developed on HRM, where Ulrich's work is particularly noteworthy, developing a kind of timeline of an HR competency model, which explains what it means to be an efficient HR

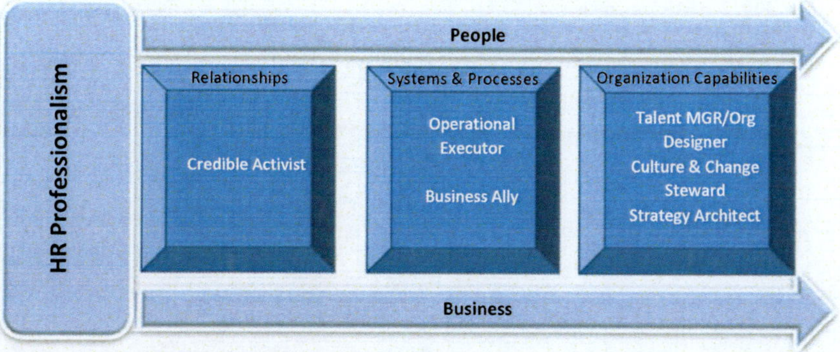

Fig. 1 HR Competency Model in 2007. *Source* Adapted from Ulrich et al. (2007, p. 6)

professional. The last updates of this model were carried out in 2012 and, with this, six collections of data were completed that trace the evolution of the competencies of an HR professional (Ulrich et al. 2012).

In the first data collection, in 1987, three important domains were found: knowledge of the business, change and delivery of HR processes. In 1992, another domain was added, that of personal credibility, and in 1997 the domain of culture was added. In 2002, the domains of change and culture were removed and the domains of HR technology and strategic contribution were inserted. In the 2007 model, illustrated in Fig. 1, domains related to relationships were found: *credible activist*; domains linked to systems and processes: *operational executor* and *business ally*; and domains linked to organizational capabilities: *talent manager/organizational designer*, *culture and change steward*, and *strategy architect*. (Ulrich et al. 2012).

This research ends up having a great impact because it defines what it means to be an efficient HR professional, that is, how to apply knowledge to changes in the business. In addition, the number of HR professionals has been steadily increasing and, therefore, it is becoming increasingly important to define what it means to be efficient in this role. In a constantly changing world, it is important to identify what HR professionals should be, know, do and deliver to contribute to the organization's success. These domains respond to skills that help answer today's business problems and issues. These domains must be updated in relation to events in the foreign market in order to transform them into internal company actions; must focus on business, people, individual skills and organizational capabilities; must include sustainable and integrative solutions; must respect the past and take into consideration the future, and contemplate long-term administrative processes and strategic practices. (Ulrich et al. 2012; see also, Ribeiro 2014; Ribeiro and Gomes 2017).

Ulrich et al. (2012) re-evaluated the HR competency model, having identified six domains that indicate the competencies needed to be an efficient professional in HRM. This study relied on focus groups, research and professional experience to identify what HR professionals should know and do. The samples included participants from Australia, Latin America, China, India, Middle East, Northern Europe and

Fig. 2 HR Competency Model in 2012. *Source* Adapted from Ulrich et al. (2012, p. 6)

South Africa and six domains were identified. The domains consisted in: *Strategic Positioner*, *Credible Activist*, *Capability Builder*, *Change Champion*, *HR Innovator and Integrator*, *Technology Proponent*, as presented in Fig. 2.

This reassessment made it possible to highlight that HRM professionals need to be activists and credible in order to build trusting relationships, and they need to have a systemic and integrated view of the business in its strategic, tactical and operational aspects. Being a credible activist is assumed as the most important competence among all (Ulrich et al. 2013). The authors also concluded that it is necessary to have a mixture of valences related to the knowledge of the external environment, the ability to build/develop capacities, support change processes, support and innovate in HR practices, and understand and use technologies. According to this study, these skills explain 42.5% of the efficiency of an HR professional. They also found that these competencies exist in various regions of the world, in different human resource roles—management, technical and administrative, as well as HRM specialists and generalists—and in all types of organizations. In addition, these skills, according to this study, explain 8.4% of business success (Ulrich et al. 2012).

This type of study also made it possible to visualize an evolution in HRM professions and, consequently, to compare the results of the various samples of participants investigated and analyzed. From 1997 to 2012, the percentage of women in the area of HR rose from 30% to 62%, while the percentage of men fell from 70% to 38% (Cabral-Cardoso 2004). Regarding academic training, it continues with a high level of degrees (51%) and, therefore, there is a greater focus on higher education for HRM professionals. It was also found that a percentage of about 25% of HRM professionals has been in positions in this area of management for less than five years. This indicates a growth in the profession and also a tendency for people to move between organizations. It is also concluded that there are about 40% to 49% of HR generalists and less specialists. It is also verified that one of the areas of greater expertise is recruitment and selection, which can mean increased attention in talent and human capital (Ulrich et al. 2013).

Finally, it is important to note that skills such as: ability to build, innovate and integrate and know how to use technologies are equally important. Furthermore, the efficiency of the professional is driven by the competence of *credible activist* and the performance of the business is increased by connecting people through technology, by aligning strategy, culture, practices and behaviors and by change sustained (Ulrich et al. 2013).

4 Analysis and Job Description and Competency Model

Next, relevant concepts for defining success profiles will be addressed, such as analysis and job description and competency model.

4.1 Job Analysis

During the time of the industrial revolution, a function was seen as something repetitive, monotonous and little stimulating, always aiming at increasing productivity. Therefore, a function can be described as the "set of tasks with one or more objectives that identify a job" (Rego et al. 2012, p.143), being it analyzed only by the functional paradigm (function as a constituent unit of the organization).

Subsequently, there began to be a shift from the functional paradigm to a competence paradigm (people as a constituent unit) making function analysis an increasingly developed area and eventually becoming a key area for HRM (Rego et al. 2012). Bilhim (2006, p.221) refers that this analysis is related to the need for the organization to know what needs to be done, even before proceeding with any recruitment and selection action. There is also the need to know what profile the future employee must have in order to successfully perform what must be done. Thus, this paradigm provides information regarding questions such as: what equipment should be used? What supervision is needed? With whom to interact with?

Thus, Brannick and Levine (2002) refer to the analysis of functions as a systematic and structured survey of the objectives, responsibilities, tasks and resources that the holders of the function have to perform in order to carry out their work within the scope of a more comprehensive profession. Saif, Khan, Rehman, Rehman, Rehman, Nawa, and Naqeeb, (2013) define the analysis of functions as the basis and an important part of HRM since it indicates what activities and accountabilities work entails, always varying from company to company.

This analysis ends up providing information that results in decision making in relation to HR planning, because it allows obtaining information about knowledge, skills and aptitudes. For example, in relation to performance evaluation, job analysis allows to identify the duties, objectives and responsibilities for each job. In relation to training and development, it helps to establish the development needs of employees and in relation to compensation, it allows defining salary structures and fair incentives and benefits. Job analysis can also influence work organization, work ergonomics, employee motivation and can be a facilitator of change (Rego et al. 2012). Hawkes and Weathington (2014) argue that job analysis serves as support in the decisions of the HRM, including those that relate the competencies with the job/processes and practices, such as recruitment and selection, training, career development and promotions. Typically, the overall job/functions analysis and description process is used in the talent attraction and retention process, and is designed to gain attention and attract candidates.

Peretti (2004) states that job analysis has the following possible applications: "recruitment and selection, staff appraisal, training of staff, preparation of career plans, definition of remuneration schedules, organization of the company and improvement of working conditions" (p. 174). Ivancevich (2008) argues that this analysis goes through several stages: examining the organization and the positions, determining the use of information from the job analysis, selecting the job to be analyzed, collecting data with the appropriate techniques, preparing the description and specification of the job.

4.2 Job Description

After the job analysis, the information collected can be recorded in various written forms, generating the description, specifications, categories and qualifications of the job. The job description turns out to be a written document about what the job holder does, why he does it, for whom he does it, how he does it, with what resources—personal and organizational—and under what conditions he does it. Usually, it includes the identification of the job, the general and specific objectives of that job, the duties and responsibilities, the materials and equipment and other conditions. The specifics of the job consist in the description of what is required in the function, that is, the demands of the function at various levels and of different nature (Rego et al. 2012).

Therefore, a job description turns out to be a dynamic reflection of the content, that is, of processes such as tasks and subtasks, the limits of competencies and responsibilities (actions initiated and decisions taken). It also reflects on the skills required to support the function, among other restricted aspects such as job change (Armstrong and Taylor 2014; Verboncu and Zeininger 2015). In summary, a job description is used to explain the roles, responsibilities, expectations and qualifications required for the jobs available as well as for those who are already occupied and which may have to be revised as a result of the changes (Armstrong and Taylor 2014; Moskwa 2016). It is a dynamic process and consequently it must be continuous and can be perceived as a factor to increase the competitiveness of strategic and operational systems and processes inside and outside organizations. It is divided into the activities that the employee must do and what he must achieve. It provides answers on what needs to be done, when, why, where, how, with whom and under what circumstances. It defines the task to be performed together with the outline of activities (Pató 2013; Rego et al. 2015).

A good job description is equally important for the organization's current employees, as it clarifies their position in the organization as part of a large team. It also provides a guide for the employee to know his possible career path. Using a consistent job description format for all jobs can ease the difficulties of creating a job description and analysis for a new position and updates to it. It also provides simple information to employees favoring the clarity of their perception of the functional role they have in and for the organization. (Moskwa 2016). This functionality is extremely important, as the knowledge and informed perception of the role played and assumed in an organization has a direct impact on the motivation and involvement of employees, and will have its repercussions on greater or lesser organizational commitment (Ribeiro and Gomes 2016, 2017).

It is also important to note that a comprehensive job description will help to systematize performance reviews and assist in the development of a work team that understands how it should fit into the overall puzzle and objectives of the organization (Moskwa 2016). Having a comprehensive and detailed job description is necessary to develop key positions and to define a clear structure (Lewis et al. 2015). It also allows contextualizing employees in the organization at the macro and micro level, as well as at the effective and affective level in terms of assuming organizational responsibilities (Ribeiro and Gomes 2016).

In addition, job description should be standardized in a language that allows comparing positions at the same level to keep all interested parties motivated, which will make the process much easier. Keep all items of the job description well structured and revised so that there are no errors in the qualification of positions and also misalignments and inconsistencies, whether due to excess or defect. The modification of a job description can generate changes in other positions; therefore, constant communication is important (Dixon 2016). It is also important to mention that each work environment is different and managers must work on an appropriate and contextualized approach to job description in each situation (Dixon 2016).

Job description can also avoid conflicts and frictions between managers and subordinates, helps with personal work, stabilizes and helps to control work achievements and can provide the basis for rationalization. It can also assist in the connection of documents within the organization and in the definition and division of tasks/responsibilities of authority/competence between companies (Pató 2013).

In a dynamic environment, a standardized job description may, however, become outdated with just a few months of work. No job can remain static, especially when competitive conditions are always changing and this change is assumed to be constant, fast and discontinuous (Ribeiro and Gomes 2016). For this reason, the job description should be reviewed at least once a year and indicate which contents are no longer relevant, what responsibilities or results need to be changed and what are the new responsibilities and results that have been added to the job. Often, the job description is just copied from another company and is focused on tasks, not processes and results, while the focus of this description should be on what the professional must do to be considered a successful professional. In addition, tasks must result in positive impacts for the company (McKenna 2015).

Having a job description makes the identification of training needs easier and, through the performance review together with the results, a more specific and relevant development plan for each practice can be created. A well-structured job description favors the removal of uncertainty and makes processes more productive, but for that to happen, rigor, communication, trust and negotiation and the quality of HRM will also be very important factors (McKenna 2015).

The effective implementation of a job analys is and description goes through several phases: definition of the strategy and objectives, planning (how to achieve the objectives and strategies, functions to be studied, schedule, techniques and legal conditions), preparation (material, environment, data collection and sorting, provisional writing of job descriptions and specifications, reviewing information with employees and definitive writing); use of the results of the analysis of functions; and update and future use (Rego et al. 2015).

4.3 *Analysis and Job Description Versus Competency Model*

It should also be noted that the analysis and job description is different from the competency models. A traditional job analysis goes through a description of the behavior, sees the job as an external object that needs to be described; the focus is on the job and in the past, the intended performance level is normal and is measured based on characteristics career-specific. A competency model considers the influence of behaviors, sees the job as a role to be disseminated, the focus is on the organization and in the future, the desired performance level is the maximum and is measured based on judgment (Sanchez and Levine 2009). However, the analysis and job description is closely linked to competence management since it ends up being a first point for the identification of competencies and, in addition, an increased number of authors suggest that the analysis and job description must be sustained, too, in what the organization intends for the future (Rego et al. 2015).

Therefore, the purpose of analysis and job description is to provide a better understanding and measure job obligations; it is linked to staffing, training and compensation. Competency models have the purpose of influencing how work tasks are performed in order to align them with the organization strategy (Sanchez and Levine 2009). In addition, a competency model must be easy to understand and communicate and is more prescriptive, serving as a driver of the organization's strategy and the behavior of its employees on a daily basis (Munck et al. 2011; Sanchez and Levine 2009).

In conclusion, analysis and job description focuses on describing and measuring job tasks while competency models create a common thread that influences employee performance in their daily lives according to the company's strategy. A job analysis aligned with a competency model will generate better results in HRM applications and practices (Sanchez and Levine 2009).

Thus, it is expected the analysis and job description, at the level of HRM, to be factual, but at the same time to present a critical perspective that allows positioning and self-evaluation. Also, aiming at the continuous improvement of professional performances and the development of organization in general, as well as each of its actors in particular (Ribeiro 2014; Ribeiro and Gomes 2016, 2017).

Then, the management of competencies is addressed, namely, its definition and some of the implications.

5 Competencies Management

It was during the 1960s that the first signs appeared that the hierarchical, rigid and stabilized structures of companies could change at any time. With external changes, organizations faced the need for restructuring in order to have greater flexibility. Thus, the competencies movement appeared in the late 1960s and early 1970s. McClelland (1973) mentioned that conventional tests of intelligence and skills were not predictable factors of performance or success in life and that they were biased for different groups. To overcome this situation McClelland (1973) proposed the term *competencies*, suggesting that these could probably provide and develop valid and impartial performance indicators. McClelland's (1973) research was carried out through interviews with professionals with high performances, identifying what they did differently to achieve that performance and using the competencies identified for selection purposes (see also, Saif et al. 2013). After this study, there was a very rapid growth in the competencies methodology. For example, Spencer and Spencer (1993) found that more than 100 researchers in 24 countries had contributed to a database with around 1000 skills in sectors, such as industry, public administration, military forces, healthcare, education and finance (see also, Cascão 2004). Also the works of Boyatzis (1982), Ulrich et al. (1995), Caldwell (2008), among others, reinforced the theoretical and practical approach to competencies in HRM.

Therefore, already at that time, competencies were highlighted in three approaches: behavioral, constructive and functional. The behavioral approach starts by analyzing the person at work, and determines, for example, by assessing critical incidents, which characteristics of the person are associated with success at work. Previously, this analysis was done with a separation between the work and the person, seeking their framing a posteriori. Spencer and Spencer (1993) argue that traditional recruitment methods do not correctly identify competencies, because many people are not well aware of their weaknesses and strengths, let alone their competencies. Therefore, what people reveal about their motivations or competencies during an interview is not enough to be credible. For this reason, a methodology was started where the candidates were asked to describe how they would behave in dealing with certain specific incidents (Ribeiro 2014; Ribeiro and Gomes 2016).

Regarding the functional approach, it is oriented towards the functions that serve the organization. It is based on knowledge, skills and attitudes, generating performance and requirements. Thus, there is a reference of a set of competencies required for the work station, linked to knowing and knowing how to do. Finally, Le Boterf (1995, 1997, 2000), in the constructive approach, argues that competence does not lie in knowledge, or skills or abilities, but in the mobilization of these particularities. Thus, the author argues that competence is not a state but rather a process of assimilation between knowing, knowing how to do and knowing how to be (Le Boterf 1995), and therefore, the combination of theoretical knowledge, operational knowledge and affective, social and cultural knowledge. Ceitil (2007) states that competencies have five components: knowing (knowledge), want-to-do (motivation), power-to-do (means and resources), know-to-do (skill and dexterity) and know-how to be (attitudes and interests). Ribeiro and Gomes (2016) add *know-how to evolve*, that is, availability and willingness to learn.

In relation, more specifically, to the concept of competence, it can be related to task and activity or as a combination of attributes underlying a good performance. This is how McClelland, in his 1973 studies, grouped competencies in five dimensions (Sousa et al. 2006):

- *Knowledge*—Knowledge acquired by an individual belonging to a specific domain;
- *Skill*—Skills that demonstrate the actual existence of acquired competencies;
- *Behaviours*—Concepts of an individual that are reflected "in their attitudes, values, emotions, actions and reactions to a situation" (p. 140);
- *Traits*—Personality traits that lead the person to behave in a certain way;
- *Motives*—Behaviors directed at something that mobilize inner forces in order to generate reactions.

However, there are authors who do not define competence in such a psychometric way, but rather as an individual characteristic that can be reflected in criteria of effectiveness and/or superior achievement in a job or situation (Boyatzis 1982; Caldwell 2008; Cascão 2004; McClelland 1973; Spencer and Spencer 1993). Spencer and Spencer (1993) address competencies such as motives, personality traits, self-concept, knowledge and skills. Since the first three are considered the center of the personality, with greater difficulty in development, the skills and knowledge are considered superficial, with greater ease for development. Thus, the "iceberg" competency model emerged, where personal self-concept, personality traits, person's motivations and self-values are considered underground skills and inputs (below-the-waterline), being related to the person's personality; skills and knowledge are considered competencies that are already observable and, therefore, outputs being demonstrated in the person's performance (McClelland 1973; Vazirani 2010; Yu et al. 2016; Zhao 2013). We are, therefore, facing a behavioral approach since the cause of an action comes from an intention, that is, a motive or a trait (Cascão 2004; Ceitil 2007; McClelland 1973; Sousa et al. 2006; Spencer and Spencer 1993; Vazirani 2010; Yu et al. 2016; Zhao 2013).

There are also other competency models, such as the functional model. This model identifies the minimum performance required for the organization to maintain the expected level of productivity and, therefore, is based on attitudes and job analysis and description. Therefore, it emphasizes specific competencies where people are considered to be stable and predictable. Another model, characterized as the Hay-McBer model, considers the distinctive competencies, that is, competencies of the best, such as creativity and innovation, aligning them with the organization's vision (Le Deist and Winterton 2005; Sousa et al. 2006).

Other authors (Saif et al. 2013) refer that a competence is the combination of knowledge (theoretical and practical), skills, abilities, values and interests. Hawkes and Weathington (2014), on the other hand, say that there is a wide variety of definitions of competencies, which makes research on competency models a little more complex.

In general, it can be concluded that competencies usually manifest themselves in acts or behaviors and are related to performance. They are observable, understandable and evaluable, they are linked to particular situations, they are considered a structured system and they are related to knowledge. However, many authors continued to consider that the concept of competence is not fully defined due to its great complexity that assumes several perspectives (Sousa et al. 2006).

Finally, it is also important to mention that with the increase and progression of globalization and multinational companies, new needs have emerged that have affected competencies. Thus, a new competence emerged: intercultural competence (Hofstede 1991, 2002; Rego and Cunha 2009). Therefore, it is important for an organization, in its internationalization process, to adapt to cultural differences and contexts. For this, it is necessary to train employees with specific technical, cultural and situational knowledge so that they develop their intercultural skills, with HRM professionals also playing a fundamental role here. In fact, cultural differences allow employees to deal with more organizational dilemmas such as, for example, universalism versus particularism, individualism versus collectivism and performance versus attributes (Ceitil 2007; see also, Hofstede 1991, 2002; Rego and Cunha 2009).

After exposing concepts such as analysis and job description and competency management, careers and talent management will be addressed, since they are concepts that also integrate the profiles of success.

6 Career Management and Talent Management

Career management is one of the biggest challenges that HRM can face today, due to the progressive changes that occur worldwide at a technological and economic level. These changes have provided an increasing complexity in modern careers and talent management is closely linked to career management. Very often individuals, who do not adapt to these new career models, end up failing in their professional progression. Thus, career management turns out to be a key to achieving personal and organizational goals (Wesarat, Majid and Sharif 2014). This management can be carried out with different practices from training to recruitment and can be guided by competency models, performance analysis, depending on the role the employee plays within the organization.

The career is linked to a process of personal and professional development throughout life. It must respond to organizational and individual needs, as both are interconnected (Rego et al. 2008). Thus, the career can be defined as: "… something that is constantly evolving, a path that is taken and that at each step, allows a different view of the surrounding environment, the path already taken and what remains to be covered (Rego et al. 2008, p. 570). Sousa et al. (2006) states that the career concept cannot focus only on the individual perspective, there must be an analysis of the relationship between the organization and the individual.

Therefore, the career is related to the *internal labor market* and there are three reasons for the existence of internal career systems for companies (Rego et al. 2008). First, economic reasons, which justify obtaining specific skills for the organization through its employees. It can cause competitive problems if these skills are transferred to other companies. Thus, it will be important to implement measures to retain these workers through, for example, career opportunities, adequate wage systems, benefits and incentives, and professional training for improvement and development (see also Ribeiro and Gomes 2016, 2017). Second, sociopsychological reasons, given that the level of loyalty and commitment of employees may be increased, if opportunities for progression and development and internal and/or external career development are created. Third, institutional and political reasons, given that there are laws and rules, companies are obliged to adopt certain practices to favor their internal customers.

However, these reasons or factors started to change due to global working conditions. Career structures, attitudes and behaviors at work have undergone changes to ensure the organization's success (Kanten et al. 2015; see also Hofstede 1991, 2002). Thus, careers that presupposed work in only one or two organizations, oriented towards safety at work and in which success would depend on promotions and status, are no longer as chosen and frequent as they used to be (Kanten et al. 2015; Rego and Cunha 2009).

Vincent (1993; see also Rego et al. 2008) suggested a type of typologies that were developed from British workers in the period between the industrial revolution and World War II. One of the typologies suggested was: the pathway of the *gold watch*, which is based on the idea that the individual remains in the same organization throughout his career. Another suggested typology was the *migration* path, which establishes that the individual makes use of a specific skill, but in different organizations. This author also suggested the *meander* path where the individual works in several companies, performing in them functions that are unrelated to each other. A final typology suggested was that of the *fracture* path in which the career path presents a discontinuity, since after a certain activity follows a completely different one (Vincent 1993; Rego et al. 2008).

Like career typologies, career choices have also been constantly changing over time. The first career choices were more linked to the importance of social structure, that is, they depended a lot on the individual's social class. Thus, career choice is dependent on the individual's own interest and is not chosen by society (Rego et al. 2008).

Today's organizations are unable to provide a lifelong job, let alone job security. As such, promotion is becoming increasingly difficult and the traditional perspective of career management is being abandoned by organizations. This led to the appearance of new forms of work such as part-time, flexible hours, among others. The new perspectives assume that each individual will have to create their level of knowledge and skills to work in different organizations and sectors. Nowadays, employees have also become more independent, they are more concerned with their satisfaction and a characteristic that they consider fundamental is continuous learning and valuation and intrinsic and extrinsic recognition (Kanten et al. 2015). Thus, careers, seen from a current perspective, end up being multiple, diversified, relational, dynamic, personalized and based on a varied path throughout life (Moen and Han 2001; Santos 2008).

Therefore, new models of individual career management and HRM practices for career development began to emerge: Boundaryless Career, Protean Career, Dual Ladder Career Path, Flexible Career, Job Posting System, Kaleidoscopic Careers, among others (Kanten et al. 2015). In the case of Protean Career, an individual career management model, individuals take responsibility for the management of their own career and seek their well-being and development. The focus turns out to be on psychological success and professional commitment, with a passage through multiple organizations. The individual assumes himself as the main responsible for his personal and professional development and growth, and manager and guide of his life (Santos 2008; Sousa et al. 2006; Rego et al. 2008).

In Boundaryless Career, another model of individual career management, there is no loyalty to the employer and inter-organizational mobility and continuous learning are the most important criteria in this type of career. Consequently, professional development is based on the ability to be an entrepreneur where a set of skills is inserted, namely: Know-Why, Know-How and Know-Whom (Kanten et al. 2015; Rego et al. 2008; Santos 2008). In Dual Ladder Career Path, an example of career development, which emerged in 1970, an individual with technical positions takes management positions at the same time. This allows opportunities for promotion and professional development (Kanten et al. 2015).

Despite all the models of individual career management, organizations have not failed to carry out career development actions. Thus, the organization develops some activities to support the success of its employees' career based on on-the-job training, performance management, assessment centers, career counseling, development programs, succession plans, mentoring, coaching, support and feedback from superiors, and measures to support work-family reconciliation (Rego et al. 2008; Wesarat et al. 2014).

It is also important to mention that career planning can also be traced from an HRM competence model that is closely related to the analysis and job description of an organization. For example, Ulrich's competency model, already mentioned, proposes the existence of individuals who make a difference, individuals who have the necessary condition and individuals who are almost catalytic levers. According to each competence, a different career plan can be built (Ulrich et al. 2007).

Another concept related to career management is talent management. The development and management of the progression of potential collaborators through various functions govern this concept. It turns out to be an effort on the part of the organization to attract, select, develop and retain key talent on a global scale so that they are aligned with the priorities of the organization's strategy (Ribeiro and Gomes 2017). This talent management is criticized because it is impossible for some organizations to take advantage of their assets without actually knowing their people and the characteristics and specificities of each one. Thus, talent management is closely linked with career management, as well as all other HRM practices (Collings et al. 2011, 2015; Ribeiro and Gomes 2016, 2017; Sanchez and Levine 2012).

Finally, after addressing concepts such as career management that are closely linked to competency management and, consequently, with the analysis and job description, the concept of success profile and its implication in organizational performance, and HRM practices, will be addressed (Sanchez and Levine 2012; Sousa et al. 2006).

7 Success Profiles

A success profile is a relatively recent concept in HRM. Consentino, Erker and Tefft (2011) developed theoretical and practical research aiming at better competence management in order to efficiently identify the competencies, personal attributes, knowledge and experiences that define and characterize an ideal performance—what they called *success profiles*. According to the authors, first it is necessary to define the most critical work factors, such as, expectations management, change management, innovation management, capacity to adapt to new global markets, and diversity management, among others. Then, success profiles are customized, identifying exactly what is needed for maximum performance. As a result, these profiles will be closely linked to the business strategy and will serve as a solid foundation for the best decisions in terms of management, use and development of individual talents. It is also possible to identify a link with the Balanced Scorecard, since this management support instrument uses the identification of critical success factors, as in the creation of profiles of success in the HRM (Sousa et al. 2006).

Therefore, these profiles are developed from:

- "What people know – technical and or/professional information needed to successfully perform job activities";

"What people have done – educational and work achievements needed to successfully perform job activities";

"What people can do – a cluster of behaviors performed on a job"; and

"What people are – personal dispositions and motivations that relate to job satisfaction, job success or failure" (Consentino et al. 2011, p. 2).

These authors propose an analysis of the success profiles concept through a holistic perspective of the processes involved upstream and downstream. Thus, the construction of these profiles will provide an articulation between individual and group strategies and the organizational strategy, ensuring the necessary adaptation and flexibility to deal with and manage very changeable and dynamic business environments. It will allow a global understanding of expectations regarding work and the self and others perception of roles and functions. It will also contribute to the recruitment and selection of candidates and boost development plans to support their success (Consentino et al. 2011).

It is also necessary to consider the relationship between success profiles and organizational performance and other HRM practices. As previously mentioned, success profiles contain critical experiences, critical knowledge, personality traits and personal values and, therefore, these success profiles will dictate the organization's performance. Thus, Child (1972) referred that the organizational structure can have an influence on performance levels and, therefore, the definition of the business context is fundamental. Other authors (Rusu et al. 2016) argue that structural factors will interfere with job performance and also clarify the company's processes from a cultural point of view. These structural factors include: organizational objectives and strategy; organization of the HR department; organizational culture; HR procedures; employee motivation and satisfaction processes and systems; and, the compensation systems. These factors contribute to the evaluation of the employees' performance, to define standards and basic performance standards, to establish standards of personalized performance and feedback regarding performance, which, consequently, will generate the potential growth of the employee's performance (Rusu et al. 2016).

In conclusion, factors in the organizational context, such as organizational objectives, strategy, technology, organizational culture, communication systems, HR principles and practices, employee motivation, leadership style, training and development processes, among others, influence the design and implementation of employee performance appraisal. Consequently, a performance assessment consistent with the organizational context represents a powerful strategy for increasing the performance of all employees (Rusu et al. 2016). Therefore, building a success profile closely linked to the organization's strategy will allow an improvement in HR practices and procedures and, predictably, an increase in organizational performance. Another study carried out (Tzabbar et al. 2017) demonstrated that most HRM practices have a positive and significant relationship in organizational performance. These practices include workplace safety practices, training, recruitment and selection, compensation practices, internal promotion, flexible work, performance evaluation, empowerment, employee involvement practices, information sharing, among others. The construction of a success profile can contribute to the development of HRM professionals—specialists and generalists—who are more focused and oriented towards the specificities and responsibilities of their professional functions. These professionals can thus contribute, within the scope of a more structured and guided HRM for the organization, for a clearer vision of the objectives and a greater organizational performance (Tzabbar et al. 2017).

Finally, other authors claim that organizational learning plays a key role in organizational success, due to rapid evolutions and changes in organizational environments (Odor 2018; Saadat and Saadat 2016). This learning is a complex, multidimensional, diverse, purposeful, dynamic, continuous, creative, persistent process, influenced by all sources of knowledge or sources of culture used by people in the organization. Although management experts say that this learning is often seen as a way of solving all organizational problems, it is a difficult process for growing and developing organizations with varying environmental conditions and in which permanent and discontinuous change requires that the learning process and systems are adequate, adaptable and adjustable, challenging and motivating (Saadat and Saadat 2016). Organizations must remain flexible and fluid, with regard to the importance they effectively attach to continuous learning, not assuming a posture of relaxation or loss of awareness and sensitivity to their surroundings, which can prevent them from seeing new challenges and new opportunities (Caldwell 2012; Odor 2018; Ribeiro and Gomes 2016, 2017; Senge 2003).

As argued by Odor (2018, p. 5), "Learning should be engrained as part of their organisation philosophy and core organisational value and culture. It is only by so doing that organisation will be able to face tomorrow when it actually comes... Finally, organisational leaders should make a gradual but holistic shift from their traditional role of figurehead, company spokesman, and resource allocator to a broader cross functional role of encouraging constructive dialogue, experimentation of ideas, which will create an environment capable of facilitating open communication". Therefore, the development of a success profile focused on experiences and knowledge, that is, on continuous learning, will enhance the effective and consequent degree of actions and decision making. (Saadat and Saadat 2016).

Finally, it is also important to add that performance evaluation turns out to be quite important since it allows a diagnosis about HR and can contribute, if perceived as fair, to improve relations with employees and increase organizational well-being as a whole (Daniels et al. 2017; Guest 2010, 2017; Kowalski and Loretto 2017; Paauwe and Boselie 2003; Peretti 2004; Ribeiro and Gomes 2016).

8 Conclusion

The overall objective of this chapter consisted of an analysis of the relevant HRM practices for building a success profile centered and contextualized in HRM professionals.

Therefore, it is important to answer to the research question: How can the elaboration of a success profile integrate the analysis and job description in order to predict a better performance of the employee and contribute to the continuous improvement of other HRM practices? In the case of a recruitment and selection process, the existence of success profiles will provide relevant information to select the candidate that is most compatible with a specific profile, increasing the likelihood that his/her performance will be in line with the organization's objectives.

Analysis and job description, competency management, career and talent management, are concepts that also integrate the concept of a success profiles. These concepts must be properly framed, built, developed in organizations, particularly and in a more consistent and systematic way in those that have a HRM department and professionals. In fact, perhaps one should start by looking for success profiles that contribute to organizations being places of greater well-being for those who work in it and who interact with it (Daniels et al. 2017; Guest 2017; Kowalski and Loretto 2017; Ribeiro and Gomes 2016).

The approach of these theoretical concepts in the literature review is a benefit, as it allows a better understanding that the construction of success profiles results from the use and articulation of HRM principles and practices and their interconnection. The systematization resulting from scientific studies facilitates a better practical understanding of the concepts, as well as greater consistency and strategic and operational awareness in the effective implementation of these concepts in the organizational scope.

In a world in which change is becoming more and more rapid and disruptive, Handy (1992), in his book *Age of Unreason*, wrote that the past is no longer a script for the future, which gives responsibility to the HRM to understand the changes that have occurred in the concept of work and worker (Cabral-Cardoso 1999; Caldwell 2003; Guest 2010). The responsibility and objective of HRM professionals, whether specialists or generalists, is to understand their roles and answer the questions: What do they do? Why do they do it? What do they do for? How do they do it? Who do they do it with? What resources do they have and what resources do they need? What do they give and receive from the organizations where they work? How do they contribute to the good performance of the organization? (Guest 2010; Legge 1995).

The ability of these professionals to adapt and adjust, to reinvent, to recreate their space for professional intervention, to relearn their functions, to identify their key competencies that contribute to more efficient and effective, responsible and ethical, sustainable and integrative decision-making, may be a current challenge that is part of a rhetoric with many decades (Legge 1995). An HRM professional who seeks continuous improvement constantly needs, in the exercise of his responsibilities and duties, to understand the evolution of contexts, the evolution and change of organizational systems and processes, and the change in the conception of work, and to know how to adjust, develop and recycle (Guest 2010; Legge 1995; Paauwe and Boselie 2003).

HRM must have a retrospective and prospective view, considering the evolution of micro and macro contexts, and cannot abdicate the perspective of systemic and critical thinking that can guarantee that it remains, and is perceived, as a guarantee of fairness and organizational equity. This will also contribute to ensuring that HRM does not lose its specificity, as a way of guaranteeing an identity that cannot be lost by the uncritical adoption of speeches and practices from other areas of management (Ribeiro 2014; Ribeiro and Gomes 2016, 2017).

References

Appelbaum, S. H., & Goransson, L. (1997). Transformational and adaptive learning within the learning organisation: A framework for research and application. *The Learning Organisation, 4*(3), 115–128.

Arditi, D., Nayak, S., & Damci, A. (2017). Effect of organizational culture on delay in construction. *International Journal of Project Management, 35*(2), 136–147.

Armstrong, M., & Taylor, S. (2014). *Armstrong's handbook of human resource management practice*. London: Kogan Page.

Barbosa, J. L. V., Kich, M. R., Barbosa, D. N. F., Klein, A. Z., & Rigo, S. J. (2015). DeCom: A model for context-aware competence management. *Computers in Industry, 72*, 27–35.

Barney, J. (1986). Organizational culture: Can it be a source of sustained competitive advantage? *The Academy of Management Review, 11*(3), 656–665.

Bilhim, J. (2006). *Gestão estratégica de recursos humanos* (2ª ed.). Lisboa: Universidade Técnica de Lisboa.

Boyatzis, R. (1982). *The competent manager—A model for effective performance*. New York: Wiley.

Cabral-Cardoso, C. (1999). Gestão de recursos humanos: Evolução do conceito, perspectivas e novos desafios. In M. P. Cunha (Ed.), *Teoria organizacional: Perspectivas e prospectivas* (pp. 225–249). Lisboa: D. Quixote.

Cabral-Cardoso, C. (2004). The evolving Portuguese model of HRM. *International Journal of Human Resource Management, 15*(6), 959–977.

Caldwell, R. (2003). The changing roles of personnel managers: Old ambiguities, new uncertainties. *Journal of Management Studies, 40*(4), 983–1004.

Caldwell, R. (2008). HR business partner competency models: Re-contextualising effectiveness. *Human Resource Management Journal, 18*(3), 275–294.

Caldwell, R. (2012). Leadership and learning: A critical reexamination of Senge's learning organization. *Systemic Practice and Action Research, 25*(1), 39–55.

Cascão, A. S. F. (2004). *Entre a gestão de competências e a gestão do conhecimento: Um estudo exploratório de inovações na gestão das pessoas*. Lisboa: Editora RH, Lda.

Ceitil, M. (2007). *Gestão e desenvolvimento de competências*. Lisboa: Edições Sílabo, Lda.

Child, J. (1972). Organizational structure, environment and performance: The role of strategic choice. *Sociology, 6*(1), 1–22.

Cohen, D. J. (2015). HR past, present and future: A call for consistent practices and a focus on competencies. *Human Resource Management Review, 25*, 205–215.

Collings, D. G., Scullion, H., & Vaiman, V. (2015). Talent management: Progress and prospects. *Human Resource Management Review, 25*(3), 1–10.

Collings, D. G., Scullion, H., & H., & Vaiman, V., (2011). European perspectives on talent management. *European Journal of International Management, 5*(5), 453–462.

Consentino, C., Erker, S., & Tefft, M. (2011). Success profiles: Comprehensive talent management throught holistic job analysis. *Development Dimensions International Inc. MMIX*. Retrieved December 14, 2020, from http://www.readbag.com/ddiworld-ddiworld-media-white-papers-successprofiles-wp-ddi.

Cunha, M., Rego, A., Cunha, R., Cabral-Cardoso, C., & Neves, P. (2016). *Manual de comportamento organizacional e gestão* (8ª ed.). Lisboa: Editora RH.

Daniels, K., Watson, D., & Gedikli, C. (2017). Well-being and the social environment of work: A systematic review of intervention studies. *International Journal of Environmental Research and Public Health, 14*(8), 1–16.

Dixon, J. (2016). United we change: Inside the art of working with unions successfully to create and revise job descriptions for the evolving needs of libraries. *Library Journal, 141*(4), 39–41.

Graen, G., Canedo, J., & Grace, M. (2020). Team coaching can enhance psychological safety and drive organizational effectiveness. *Organizational Dynamics, 49*(2), 1–6.

Guest, D. (2010). Human resource management and performance: Still searching for some answers. *Human Resource Management Journal, 21*(1), 3–13.

Guest, D. E. (2017). Human resource management and employee well-being: Towards a new analytic framework. *Human Resource Management Journal, 27*(1), 22–38.
Handy, C. (1992). *A era da irracionalidade*. Cetop: Lisboa.
Hassan, S. (2016). Impact of HRM practices on employee's performance. *International Journal of Academic Research in Accounting, Finance and Management Sciences, 1*(6), 15–22.
Hawkes, C., & Weathington, B. (2014). Competency-based versus task-based job descriptions: Effects on applicant attraction. *Journal of Behavioral and Applied Management, 15*(3), 190–221.
Hofstede, G. (1991). *Cultures and organizations-software of the mind*. London: McGraw-Hill.
Hofstede, G. (2002). Dimensions do not exist: A reply to Brendan McSweeney. *Human Relations, 55*(11), 1355–1361.
Huselid, M., Jackson, S., & Schuler, R. (1997). Technical and strategic human resource management effectiveness as determinants of firm performance. *Academy of Management Journal, 40*(1), 171–188.
Ivancevich, J. (2008). *Gestão de recursos humanos* (10ª ed.). São Paulo: McGrawHill.
Jaiswal, P. R. (2018). Human resource management and its importance for today's organizations. *International Journal of Advance and Innovative Research, 1*(5), 28–36.
Jones, G. (2010). *Teoria das organizações* (6ª ed.). São Paulo: Pearson Education do Brasil Ltda.
Kanten, P., Kanten, S., & Yesiltas, M. (2015). New perspectives in career management: boundaryless career and dual ladder career path. In *Proceedings of the Multidisciplinary Academic Conference* (pp. 319–327).
Kapoutsis, I., Papalexandris, A., & Thanos, I. (2019). Hard, soft or ambidextrous? Which influence style promotes managers' task performance and the role of political skill. *The International Journal of Human Resource Management, 4*(30), 618–647.
Kaufman, B. E. (2001). The theory and practice of strategic HRM and participative management antecedents in early industrial relations. *Human Resource Management Review, 11*(4), 505–533.
Kowalski, T. H. P., & Loretto, W. (2017). Well-being and HRM in the changing workplace. *The international Journal of Human Resource Management, 28*(16), 2229–2255.
Landy, F. J., & Vasey, J. (1991). Job analysis: The composition of SME samples. *Personnel Psychology, 44,* 27–50.
Le Boterf, G. (1995). *De la compétence, essai sur un attracteur étrange*. Paris: Editions d'Organisations.
Le Boterf, G. (1997). *Compétence et navigation professionnelle*. Paris: Editions d'Organisation.
Le Boterf, G. (2000). *Construire les compétences individuelles et collectives*. Paris: Editions d'Organisations.
Le Deist, F. D., & Winterton, J. (2005). What is competence? *Human Resource Development International, 8*(1), 27–46.
Legge, K. (1995). *Human resources management: Rhetorics and realities*. London: The Macmillan Press.
Lisboa, J., Coelho, A., Coelho, F., & Almeida, F. (2004). *Introdução à gestão de organizações*. Porto: Grupo Editorial Vida Económica.
Markoulli, M. P., Lee, C. I. S. G., Byington, E., & Felps, W. A. (2017). Mapping human resource management: Reviewing the field and charting future directions. *Human Resource Management Review, 27,* 367–396.
McClelland, D. C. (1973). Testing for competence rather than for "intelligence". *American Psychologist, 28*(1), 1–14.
McKenna, P. (2015). A key performance issue: When job descriptions don't do the job. *Of Counsel, 34*(5), 15–18.
Melo, P., & Machado, C. (2015). *Gestão de recursos humanos nas Pequenas e médias empresas: Contextos, métodos e aplicações*. Lisboa: Rh Editora.
Mitsakis, F. (2014). Human resources (HR) as a strategic business partner: Value creation and risk reduction capacity. *International Journal of Human Resource Studies, 4*(1), 154–170.

Moen, P., & Han, S.-K. (2001). Gendered careers: A life-course perspective. In R. Hertz & N. L. Marshall (Eds.), *Working families: The transformation of the American home* (pp. 42–57). Berkeley: University of California Press.

Moskwa, L. (2016). Do your firm's job descriptions and performance reviews meet the standard? *Journal of Pension Benefits: Issues in Administration, 23*(2), 53–55.

Munck, L., Souza, R. B., Castro, A. L., & Zagui, C. (2011). Modelos de gestão de competências versus processo de validação. Um ponto cego? *Revista de Administração, 46*(2), 107–121.

Odor, H. (2018). A literature review on organizational learning and learning organizations. *International Journal of Economics & Management Sciences, 7*(1), 2–6.

Paauwe, J., & Boselie, P. (2003). Challenging strategic human resources management and the relevance of the institutional setting. *Human Resource Management Journal, 13*(3), 56–70.

Parmenter, D. (2010). *Key performance indicators: Developing, implementing and using winning KPIs* (2ª ed.). Hoboken, New Jersey: John Wiley & Sons Inc.

Pató, B. (2013). The 7 most important criterions of job descriptions. *International Journal of Business Insights & Transformation, 7*(1): 68–73).

Peretti, J. M. (2004). *Recursos humanos* (3ª ed.). Lisboa: Edições Sílabo.

Rego, A., & Cunha, M. (2009). *Manual de gestão transcultural de recursos humanos* (1ª ed.). Lisboa: Editora RH.

Rego, A., Cunha, M., Gomes, J., Cunha, R., Cabral-Cardoso, C., & Marques, C. A. (2008). *Manual de gestão de pessoas e do capital humano* (2ª ed.). Lisboa: Edições Sílabo.

Rego, A., Cunha, M., Gomes, J., Cunha, R., Cabral-Cardoso, C., & Marques, C. A. (2012). *Manual de gestão de pessoas e do capital humano* (3ª ed.). Lisboa: Edições Sílabo.

Rego, A., Pina e Cunha, M., Gomes, J., Campos e Cunha, R., Cabral-Cardoso, C., & Marques, C. (2015). *Manual de gestão de pessoas e do capital humano*. Lisboa: Edições Sílabo.

Ribeiro, J. L., & Gomes, D. (2017). What is talent management? The perception from International Human Resources Management students. In C. Machado (Ed.), *Competencies and (global) talent management*. New York: Springer.

Ribeiro, J. L. (2014). *A percepção social acerca da gestão e dos gestores de recursos humanos. Uma análise enraizada nos dados*. Tese de Doutoramento em Ciências Empresariais. Braga: Universidade do Minho.

Ribeiro, J. L., & Gomes, D. (2016). Other organizational perspectives on the contribution of human resources management to organizational performance. In C. Machado & J. P. Davim (Eds.), *Organizational management: Policies and practices in a global market* (pp. 63–106). Londres: Palgrave Macmillan.

Rusu, G., Avasilcai, S., & Hutu, C. (2016). Organizational context factors influencing employee performance appraisal: A research framework. Procedia—*Social and Behavioral Sciences, 221*, 57–65.

Saadat, V., & Saadat, Z. (2016). Organizational learning as a key role of organizational success. *Procedia—Social and Behavioral Sciences, 230*, 219–225.

Saif, N., Khan, M., Rehman, K., Rehman, S., Rehman, Z., Nawa, T., et al. (2013). Competency based job analysis. *Internacional Journal of Academic Research in Accounting, Finance and Management Sciences, 3*(1), 105–111.

Sanchez, J., & Levine, E. (2009). What is (or should be) the difference between competency modeling and traditional job analysis? *Human Resource Management Review, 19*(2), 53–63.

Sanchez, J., & Levine, E. (2012). The rise and fall of job analysis and the future of work analysis. *Annual Review of Psychology, 63*, 397–425.

Sanders, K., Shipton, H., & Gomes, J. (2014). Is HRM process important? Past, current and future challenges. *Human Resource Management, 53*(4), 489–503.

Santos, G. (2008). *O desenvolvimento de carreira dos académicos: Uma análise centrada na relação entre o trabalho e a família. Tese de Doutoramento.* Braga: Universidade do Minho.

Senge, P. (2003). Taking personal change seriously: The impact of organisational learning on management practice. *Academy of Management Executive, 17*(2), 47–50.

Sousa, M., Duarte, T., Sanches, P., & Gomes, J. (2006). *Gestão de recursos humanos: Métodos e práticas.* (7ª Ed.). Lisboa: Lidel—Edições Técnicas, Lda.
Spencer, L. M., Jr., & Spencer, S. M. (1993). *Competence at work: Models for superior performance.* New York: Wiley.
Tornow, W. (1984). The codification project and its importance to professionalism. *Personnel Administrator, 29*(6), 84–86.
Tzabbar, D., Tzafrir, S., & Baruch, Y. (2017). A bridge over troubled water: Replication, integration and extension of the relationship between HRM practices and organizational performance using moderating meta-analysis. *Human Resource Management Review, 27*(1), 134–148.
Ulrich, D., & Dulebohn, J. H. (2015). Are we there yet? What's next for HR? *Human Resource Management Review, 25*(2), 188–204.
Ulrich, D., Brockbank, W., Johnson, D., & Younger, J. (2007). Human resource competencies: Responding to increased expectations. *Employment Relations Today, 34*(3), 1–12.
Ulrich, D., Brockbank, W., Young, A., & Lake, D. (1995). Human resource competencies: An empirical assessment. *Human Resource Management, 34*(4), 473–95.
Ulrich, D., Younger, J., Brockbank, W., & Ulrich, M. (2012). *The new HR competencies—Business partnering from the outside-in.* Boston: The RBL Group.
Ulrich, D., Younger, J., Brockbank, W., & Ulrich, M. (2013). The state of the HR profession. *Human Resource Management, 52*(3), 457–471.
Vazirani, N. (2010). Review paper: Competencies and competency model—A brief overview of its development and application. *SIES Journal of Management, 7*(1), 121–131.
Verboncu, I., & Zeininger, L. (2015). The manager and the managerial tools: Job description. *Review of International Comparative Management, 16*(5), 603–614.
Vincent, D. (1993). Mobility, bureaucracy and careers in twentieth-century Britain. In A. Miles & D. Vincent (Eds.), *Building European society, occupational change and social mobility in Europe, 1840-1940* (pp. 217–239). Manchester: Manchester University Press.
Wesarat, P., Majid, A., & Sharif, M. (2014). A review of organizational and individual career management: A dual perspective. *International Journal of Human Resource Studies, 4*(1), 101–113.
Yu, Y., Sheng, J., & He, B. (2016). College students' professional ability training based on the theory of competency. In *International Conference on Education, E-learning and Management Technology* (EEMT), Xi´an, China.
Zhao, L. (2013). The research on the performance evaluation of civil servant based on competency model. In *International Conference on Artificial Intelligence and Software Engineering* (ICAISE), Nanjing, China.

The Misconception Between Organizational Learning and Knowledge Management

Diana Fernandes and Carolina Feliciana Machado

Abstract The present work assumes the concepts of organizational learning and knowledge management are related, although not identical. Through literature review, we provide clarification on the specificities of each conceptual construct: even if the basic processes and objectives of knowledge management and organizational learning are mutual, we contend that knowledge management refers to a conceptual construct which primarily focuses on knowledge content, while organizational learning entails all aspects of data, information and knowledge. So, clear differences can be presented when comparatively analysing such conceptual constructs, which we summarized into five distinctive core vectors, operating at the level of the conceptual approach (interest of academics vs. interest of practitioners), the goal (process vs. result), the scope (system vs. subsystem), the agents (individual level vs. collective level) and the products (individual memory vs. organizational memory). Nonetheless, based on the common elements vividly detected, shared purposes are identified, once such notions jointly thrive towards enhanced information and knowledge that allow organizational behaviours, impelling on superior organizational performance, objectives which guide the solidification and the expansion of a knowledge value chain model able to incorporate and objectivize the basis for a framework towards the progress in knowledge management programs and in the enhancement of a learning organization.

1 Introduction

With the advent of Globalization, we enter a new era marked by powerful forces that impact all aspects of life, at individual, societal and organizational level. The triumph of the global economy and financial markets, alongside with the banalization

D. Fernandes · C. F. Machado (✉)
School of Economics and Management, University of Minho, Braga, Portugal
e-mail: carolina@eeg.uminho.pt

C. F. Machado
Interdisciplinary Centre of Social Sciences (CICS.NOVA.UMinho), University of Minho, Braga, Portugal

C. Machado and J. P. Davim (eds.), *Knowledge Management and Learning Organizations*,
https://doi.org/10.1007/978-3-030-71079-8_7

of mass media and technology (with particular highlight to the Internet), operated a radical transformation of the work environment, increasing customer power and opening the path the to the rise of knowledge and learning as chief organizational assets. Henceforth, the intellectual component of products and services has been progressively conquering significance, which consubstantiated knowledge management development as a key organizational pillar: indeed, in the era of knowledge, efficient organizations are the ones which learn, memorize and act on the basis of the available information and knowledge, on the most optimized way (Davenport and Prusak 1998).

Thus, in the current era we assist to a growing workplace diversity and mobility, hence, employees' roles and expectations have been reorganizing in face of the new environmental configuration. This shift then recommends the relevance of assessing and evaluating progress in knowledge management and organizational learning activities. The basis for competition during the last decade then started to shift, so that in the current era we assist to a focus on knowledge (its creation, transformation, appropriation, diffusion and management) as a source of organizational competitive advantage, providing valuable inputs towards the increase of employee and customer satisfaction, increase of profits, efficiency improvement and lower operational costs (Wah 1999; Davenport et al. 1998).

Based on that, knowledge management and the learning organization have been assuming as two of the potentially most important concepts in the corporate new era, due to their input towards allowing organizations to transform themselves to be competitive in the new millennium (Sethi and King 1994).

Even though the potential significance of knowledge management and the learning organization as conceptual and managerial approaches has been widely understood and recognized, both in academical debates as well as in organizational processes and practices, tested and recognized empirical methodologies for assessing progress towards the allied goals of effective knowledge management and/or the creation of a learning organization are not yet developed (Davenport et al. 1998; Sveiby 1997; DiBella et al. 1996; Schein 1993). Indeed, the two areas of knowledge management and organizational learning have been used to define one another, often overlapping despite conceptual idiosyncrasies: reasons ground precisely on the misconception between both theoretical constructs, so that thinking and empirical analysis then becomes ambiguous, lacking rigour, thus hampering benefits to be achieved, tracked, understood and measured.

Hence, a learning organization can be defined as an organization capable of efficiently generating, attaining and transferring knowledge, thus able to modify its behaviour on a structured, organized and constant basis to reflect further new knowledge, capitalizing on it (Garvin 1998). Indeed, it consists of an organization where employees excel at creating, acquiring, assimilating, transferring and transforming knowledge, at a constant rhythm all along the organizational structure and agents. Thus, three building blocks reside at its core, namely institutions including a supportive learning environment, established learning processes and practices, and leadership behaviours that strengthen learning as a required process in order to improve both individual as well as organizational performance (Garvin et al.

2008). So, the learning organization conceptual construct focuses on an organization as an entity, by its side, organizational learning concentrates on the learning process within the organization (Sicilia et al. 2006; Yeo 2005; Loermans 2002). Hence, both conceptual constructs are concerned with processes for obtaining information, interpreting data, developing knowledge and sustaining learning (Kezar 2005). On the other hand, an organization acknowledges that managing its knowledge the most optimized way possible is crucial for organizational development and to leverage organizational performance (Blodgood and Salisbury 2005), so that knowledge management assumes pertinence in management debates as it relates to the above mentioned conceptual constructs given the fact that it includes activities such as generating, organising, storing, sharing, using and transforming knowledge (Wong and Aspinwall 2004).

Therefore, the present work is based on the premise that the two concepts (organizational learning and knowledge management) are obviously related, with clear commonalities, although not identical: there are conceptual differences to highlight, which then translate into different managerial applications and outcomes. This is so given the fact that both concepts deal with the same content constructs: data, information and knowledge, both concentrate on processes architected at acquiring, refining, storing, using, transmitting and transforming the knowledge content within an organizational context, hence, both as well partage the goal of leveraging business performance through these processes.

Thus, through literature review, in the present work we aim to fill in this gap of conceptual misconception, providing further clarification on the specificities of each conceptual construct, presenting the respective background, main characteristics and operative vectors of each notion, thus exposing and elaborating on the aspects in which they relate and differentiate. As so, it organizes and articulates into five sections: first, we provide an initial background contextualizing the relevance of knowledge in the organisational context; second, we elaborate on the concepts of learning organization and organizational learning; the same being done on the third section but this time in regards to the concept of knowledge management; in section four we systematize and fundament the core differences between such conceptual constructs and, lastly, in section five we address the relevance of learning and knowledge in the organizational context as a means to leverage organizational performance. The present work then consubstantiates as a theoretical contribution aimed at providing more rigour to conceptual debates around the notions of learning organization, organizational learning and knowledge management.

Although the basic processes and objectives of knowledge management and organizational learning are mutual, we contend that knowledge management refers to a conceptual construct which primarily focuses on knowledge content, while organizational learning entails all aspects of data, information and knowledge. So, clear differences can be presented when comparatively analysing such conceptual constructs, which we summarized into five distinctive core vectors, operating at the level of the conceptual approach (interest of academics vs. interest of practitioners), the goal (process vs. result), the scope (system vs. subsystem), the agents (individual level vs. collective level) and the products (individual memory vs. organizational memory).

Nonetheless, based on the common elements vividly detected, communal purposes are identified, once such notions jointly thrive towards enhanced information and knowledge that allow organizational behaviours, impelling on superior organizational performance, objectives which guide the solidification and the expansion of a knowledge value chain model able to incorporate and objectivize the basis for a framework towards the progress in knowledge management programs and in the enhancement of a learning organization.

2 Knowledge in the Organisational Context

Drucker (1992) set forward that the significance of knowledge could be originating economic benefits to organizations as well as to the society, thus presenting the concept of "knowledge economy" to explain the current worldwide economic structure and dynamics. Therefore, we can clearly observe a paradigm shift towards knowledge as the core source of value, crystalizing that new economy should be led by the people who can effectively and efficiently create, retain, transform, integrate, share and manage knowledge between themselves and throughout organizational dynamics.

Knowledge is currently the ultimate organizational competitive advantage, thus having gathered increasing managerial focus. Therefore, once it will frame the argumentation throughout this article, it is essential to at first clarify this concept, foremost due to the interconnection, in theoretical as well as in practical terms, of the concepts data, information and knowledge, which, nevertheless, does not hinder the critical distinctions that can be made between all of them (Spiegler 2000; Tuomi 2000). Indeed, not only are the definitions of the three units distorted, in fact the literature (e.g. Choo et al. 2007; Spiegler 2000; Davenport and Prusak 1998; Ackoff 1997) often describes such relationships as asymmetrical, via a rigid hierarchy where data resides at the bottom, followed by information climaxing with knowledge.

Due to the crystallization of the Globalization, information is increasingly available to everyone, so that it can be understood as processed data, able to be stored within computers, books, etc. (Harari 1997). Again, with the advent of Globalization, digital formats of information storage have emerged and triumphed in the current era. Subsequently, we can defend that information refers to a broader concept that overcomes data, as it refers to a set of data that is organized and structured within a context, based on the premise that data lacking any structure and organized course is meaningless and purposeless. On the other hand, the process of cognizance, understanding and acquaintance we can accomplish through observation, investigation, examination, study and/or experience via a systematic and organized path over the course of time is commonly defined as knowledge. So, it becomes clear that knowledge is a broader and a refined concept, in the sense that we can say that humans inherently possess and process knowledge, based on their process of information appropriation (Malhotra 1998). Hence, this argument then develops to defend that knowledge can be conceived as an individual's interpretation of information, it being grounded on his/her personal experiences, life background, skills and competencies: thus, in a traditional viewpoint we can argue that knowledge consists of a broader notion that overcomes information, as it refers to the process of gathering, processing, appropriating and assimilating a set of information through a process which aims at providing added value.

Based on this, we can be aware of the critical distinctions between both concepts even though also evidently report its commonalities and interconnectedness. To Davenport et al. (1998), knowledge is a decidedly valuable resource which originates from a combination of context, experience, interpretation, reflection and information. Hence, knowledge production can be perceived as a process embracing four sub-processes, namely information acquisition, individual and group learning, knowledge claim formulation and knowledge claim evaluation. By its end, and furthering this dynamics, knowledge integration comprises other four sub-processes, which are knowledge and information broadcasting, searching/retrieving, knowledge sharing and teaching, acknowledging that interpersonal, electronic or both methods are possible to be used in the execution of such sub-processes. Subsequently, we can state that knowledge processes culminate into knowledge outcomes, based on the premise that knowledge consists in an encoded, tested and still surviving structure of information that helps the agent, its creator, to constantly adapt to the system. Hence, we can suggest that data and information are two contrary ends on a continuum, the influential factor will then be the knowledge the beholder brings into the situation. Indeed, information is regarded as the fundamental element in knowledge organisation (Choo 1998) as it is the basis for the production of human and organisational knowledge. This line of thinking is set forward by Spiegler (2000), who argues that the Triade data-information-knowledge are concepts that, though distinct, relate through a circular dynamic so that knowledge eventually reverts back to data: so, the Triade operates in a symbiosis dynamic, via an ongoing circle. Following the same logics, Tuomi (2000) also argues that, given the fact that knowledge is needed to know which data to obtain, knowledge management thus concerns, in its very essence, to the knowledge-information-data sequence.

This argument can then be developed so that we are able to defend it to be consistent with that of explicit knowledge converted into tacit knowledge (Nonaka 1994). These are the two types of knowledge: explicit knowledge is plainly and evidently formulated, expressed or defined, hence easily articulated without ambiguity, being codified and stored in a database; on the other hand, tacit knowledge consists on the unarticulated knowledge residing in a person's brain, therefore frequently difficult to verbalize, codify, transfer and appropriate. It includes lessons learned through the life experience, judgment, know-how and even intuition (O'Dell et al. 1998). So, linked to the stated above, following the argumentation of Nonaka (1994) we can advocate that the process of searching, observing and being aware of certain aspects may be defined as a collection of data which will serve as information when provided some meaning, thus setting forward the argument which presents the Triade data-information-knowledge working as a symbiosis. A slight variation can be presented, if we consider the process of searching for a precise and explicit set of data and/or information based on existing knowledge, thus setting forward as well the argument which presents the Triade knowledge-information-data argument working as a symbiosis.

Drifting the analysis to the organizational context, we advocate that organizational knowledge is conceived to refer to what organizational actors know about their organization, know about themselves as organizational actors and know about organizational dynamics, focusing on competitors, customers, products, services, processes, errors and achievements (O'Dell et al. 1998). Therefore, knowledge has wide-ranging impacts on the organization, so, considering the aspects inherent to the organisational context, knowledge must involve the act of knowing transfigured into the problems and in efficient practices of individuals and organizational groups (Spender 2006). Consequently, and alluding to Wiig et al. (1997) and to Blackler (1995), it is important to notice that knowledge has valuable idiosyncrasies as it is unstable, ephemerous, volatile, intangible, problematic to access quantitatively, possible to be simultaneously used by different processes, thus also increasing and turning more profound in its broadness with use, as well being habitually personified and stored in agents with expectations and wills, personal background and different life paths, hence can be manifested and applied to distinct aims. This way, knowledge in the organizational context is stored and organized in databases and/or is spread and appropriated through sharing of experiences, know-how and best practices (we can as well in these dynamics have the input of other both internal and external sources to the organization): thus, it is important to notice that the sharing of tacit knowledge is a vital characteristic of team-based learning organizations: as it accumulates over time through a sustained, organized and long-term process, it fosters the achievement of corporate acute understanding because organizations are then able to gather further knowledge and accomplish efficiency, through such continual and hampered expertise.

Being a strategic asset, we can support that organizations that intend to persist competitive should foster mechanisms for capturing relevant knowledge and disseminating it accurately, reliably, concisely and in a timely manner to all agents in possible need of it. So, at an organizational level, knowledge is able to provide a valuable source of competitive advantage (Hendricks and Vrien 1999; Nonaka and Takeuchi 1995), as it possesses four key characteristics (Aggestam 2006). It is inimitable, given the fact that each individual member of the organization actively contributes to knowledge creation, transformation, application and diffusion, based on his/her personal understanding of information. Subsequently, as individuals aggregate in teams/departments at the organizational structure, assimilation of knowledge is also processed through group interpretations, thus is dependent on the synergy of the total members and their membership dynamics. This resource is as well inimitable because organizational knowledge is constructed rooted on the exclusive history of the organization, thus erects on the organization's experiences and accumulated expertise. This argument can develop into another, characterising knowledge by its rareness, based on the fact that organizational knowledge is, at its very core, reliant on the knowledge and experiences of actual and former employees, hence constructed on specific organizational capital prior to knowledge. Following this argument, we can characterize knowledge by being non-substitutable, because the synergy of specific groups cannot be replicated on its exact terms: consequently, the group embodies idiosyncratic skills and capabilities which are non-substitutable. We then state that knowledge is as well valuable, helping the organization to attain a chief strategic advantage, once new organizational knowledge results in improved approaches, processes, practices, technologies, products and/or services, subsequently impelling the organization to remain competitive and viable in the long term.

Based on the above, we can then elaborate on the ideal context for organisational knowledge creation, which we sustain to be organised in three dimensions, following the line of Nonaka et al. (1996). At first, we need to shed light on the environments and on the dynamics of the observed relationships, keeping in mind the advantages of disposing of a solicitude environment, marked by empathy, confidence, trust, understanding, accessibility, integrity, stimulus to courage and constant help. In such environments, it is also chief to dispose of an adequate level of informality so that the organizational actors can perceive a freedom atmosphere which enables them to boost interactive relationships and leverage their openness to knowledge sharing. Then, attention shall be drifted towards structures, which in an ideal context for organizational knowledge shall dispose of a communication infrastructure hampering and facilitating the flow of information and sharing: thus, this ideal structures shall be horizontal, entailing few and soft hierarchical levels, without strict barriers derived from tenure or status. Managerial policies and actions also deserve a sharp focus, based on the premise that an ideal context for organizational knowledge shall foster the appropriate conditions for sharing perspectives, visions, aims and strategies, which means, it is vital that such policies and actions actively stimulate taking risks and the feeling of compassion and understanding in face of errors, as so boosting individual, team and organizational resilience. It will be attained by encouraging and valuing diversity and systemic thinking, through which organizations will be able to objectivize the promotion of a fluid and ongoing stream of ideas, adopting flexible policies and implementing an adequate autonomy level to operationalize a constant promotion of information propagation. As well, such ideal managerial policies and actions shall foster the most efficient tools and practices to create and implement adequate procedures for knowledge creation processes, focusing as well on providing group learning conditions that encourage knowledge creation and diffusion, disseminating knowledge to all organisational levels, based on the previous broader understanding of the required knowledge to achieve the defined aims.

Furthering the above, in order to explain organisational knowledge development Nonaka and Takeuchi (1995) propose that the collaborative interface between the individuals' tacit and explicit knowledge consists of the key dynamics in the organisational knowledge creation process, which is composed of four modes (socialisation, externalisation, combination and internalisation) from whose interaction knowledge arises, via a spiral path.

3 Learning Organization and Organizational Learning

In the current highly volatile business environment, the recurring transformation of organizations from a resource-based paradigm towards a knowledge-based one has decisively protracted the importance of organizational learning, subsequently, knowledge and organizational learning have been assuming as sources of competitive advantage for the contemporary organizations (Liao and Wu 2010; Chuang 2004). This is grounded on the premise that organizations that dispose of a clear, structured, organized and continuous process concerning organizational learning can address and manage organizational issues, namely organizational regeneration, transformation and innovation, through a more efficient, accurate, clear and fast process, thus sustaining their competitive advantage sources and, subsequently, their position in the market (Bierly et al. 2000; Goh and Richards 1997).

Leading in this reflection is the concept of learning, which by its essence refers to the process dynamized towards the achievement of knowledge, capacities and skills from an agent, through studying and understanding. Therefore, it is clear that the vital part of the learning process corresponds to the experience, reflection, theorization and experimentation, core stages that, even though being distinct, operate in a cadence, articulating a symbiosis which then condenses into a complex, dense and intricated process, as these elements depend on each other. Relevant to notice is that, underlying this definition, there is the assumption that all organizational agents understand internal and external aspects of learning (Kolb 1984).

The analysis of organizational learning has recently conquered attention both in academical debates and managerial approaches, once it has been conceived as one of the key strategic vectors to accomplish long-term organizational success (Cunningham and Gerrard 2000; Harung 1996; Senge 1990). Indeed, empirical research has explained and reported the importance of organizational learning focusing on its positive impacts towards innovation leveraging, problem-solving, effective implementation of change, strategy formulation and execution, organizational renewal and revival, achievement of competitive advantage and performance outcomes (Adams and Lamont 2003; Bierly et al. 2000; Edmondson and Moingeon 1998; Bass and Avolio 1993; Attewell 1992).

The notion that an organization could learn and capitalize on such knowledge was a key advance, first articulated by Cyert and March (1963) who presented a theory of organizational learning as a component of an organizational decision making model, rooted on the role of rules, procedures and routines in response to external threats. Remarkable contributes are the idea that it is via an organizational learning process that the organization itself adjusts to its environment, learning from its experience and capitalizing from it in the long-run, more efficiently addressing responses to environmental changes. Argyris and Schön (1978) and Cangelosi and William (1965) foster the notion of organizational learning as a discontinuous process, offering a model based on tensions between individual and organizational levels of learning, criticizing the assumptions of Cyert and March (1963) by stressing that human behaviour within organizations frequently is not bonded by economic rationality, based on the argument that individuals and organizations often seek to protect themselves from the setbacks of the learning experience by establishing defensive routines. Later, Fiol and Lyles (1985), Daft and Weick (1984) and Hedberg (1981) offered important contributions to terminology acuteness and to deeper perspectives on organizational learning. Epple et al. (1991), Huber (1991), March (1991) and Simon (1991) have, later on, been very influential once they recognized there are substantial antecedents in the organizational learning process (mainly human), as so promoting it is desirable to maximize the efficient use of knowledge in organizations. Though profiting from such inputs, the idea of a learning organization has a more recent derivation, evolving by the end of the 1980s largely benefiting from the European work, specifically by Pedler et al. (1989) and Garratt (1987). Senge's (1990) contribution is commonly emphasised as foundational, nevertheless, critical work (e.g. Coopey and Burgoyne 2000; Snell and Chak 1998; Coopey 1995) raised concerns about such notion of a learning organization: not necessarily undermining it, they questioned its cultural limits (stating it was not necessarily transferable to other cultural contexts), labelling it as politically naïve and embracing an ideology that was exploitative of employees. Later authors shared this line of thinking, developing it by stressing that theories and practices around learning organizations should include core concepts such as power, politics and culture (Lawrence et al. 2005) and evidently elaborate on the impact towards business success (Thomas and Allen 2006).

Hence, a learning organization can be defined as an organization capable of efficiently generating, attaining and transferring knowledge, thus able to modify its behaviour on a structured, organized and constant basis to reflect further new knowledge, capitalizing on it (Garvin 1998). Indeed, a learning organization consists of a place where employees excel at creating, acquiring, assimilating, transferring and transforming knowledge, at a constant rhythm, all along the organizational structure and agents. Thus, at the core of this conceptual construct we observe three building blocks, namely institutions including a supportive learning environment, established learning processes and practices, and leadership behaviours that strengthen learning as a required process in order to improve both individual as well as organizational performance (Garvin et al. 2008).

By valuing this process of continuous improvement, a learning organization can define with more accuracy and efficiency its aims, designing strategies and processes through systematically identifying the necessary steps that comprise the path to reach such goals, using the principles and practices of continuous learning. Henceforth, we can postulate that a learning organization at its essence refers to a system which determines the errors and corrects it, on a constant basis and aiming at the long-term organizational viability, thus, this conceptual construct aims to augment and to optimize the organizational adaptability to change (Jamali and Sidani 2008).

As so, Senge (1990) defines a learning organization highlighting its process of continuing expansion, aiming at creating its future via a sustained and enduring basis. Pedlar et al. (1991) share this perspective, sustaining that a learning organization demands that an organization makes a sentient effort to allow learning activities, to generate adaptation and to boost development abilities all along its structures, processes, practices, products and services, through a structured, sustained and enduring path throughout their life. On the same line, Thomas and Allen (2006), Gopher et al. (1994), Solomon (1994), and Thornburg (1994) described that a learning organization is one that possesses an enhanced capacity to learn, adapt and change, as a condition to it enabling employees to consistently acquire and share knowledge. Garvin (1993), however, impels Senge's (1990) viewpoint to a more rigorous approach, claiming it vagueness. Subsequently, conceives a learning organization as one skilled at creating, achieving and shifting knowledge, aimed at a continuous process of improving its behaviour to reflect new knowledge through renewed tools, practices and processes. To Chinowsky and Carrillo (2007), a learning organization is skilled at originating, achieving, transferring and applying knowledge, hence embracing change and innovation all along its structure, which fosters optimum performance and maximum competitive advantage. Definitely, we can state that essentially the learning organization orients its strategies and actions towards the future, because it attempts to reveal and understand the underlying causes of events in order to solve problems effectively, this way remaining vividly opened to learn from past events and mistakes. Therefore, considers long-term approaches rather than focusing on the present and short-term goals (Müller and Stocker 2011). Henceforth, the concept of learning organization perceives the organization as a unified entity, focussing on its characteristics that reassure its members are able to learn. Garvin (1993) then climaxes by suggesting that a more specific definition of a learning organization is required so that managers are able to generate value from this approach.

As a summary of the above conceptual proposals, we can state that the overall definition of a learning organization is constructed on five characteristics. Based on the premise that individuals can be regarded as core subsystems in the organization, we can advocate that organizational learning starts at the individual level, as it ground resides in individual learning. So, it is considered to depend on the collective cognitive processes of individuals collectively structuring the organization (Yeo 2005). This is to say, organizations possess organizational learning capabilities underlying individual learning (Nonaka and Takeuchi 1995). Nonetheless, individual learning does not unavoidably imply changes in organizational knowledge, given the fact that organizational knowledge is indeed independent of specific organizational members, as for example the knowledge derived from knowledge repositories and embedded in policies and routines. So, we can see that rooting this concept there is a strong focus on the individual, nevertheless it does not extinguishes at that level, based on the premise that a learning organization is structured and organized to operationalize the scanning for information in its environment, creating as well information by itself and heartening its members, at the individual level, to connect in order to transfer knowledge between themselves as a team (Jensen 2005). As so, leadership, understood as the ability to lead the organization towards the implementation of a learning organization, is a chief aspect of a learning organization, following the assumption that a strategic leadership in a learning organization is the motor to impel that work processes integrate every aspect of knowledge. So, such leadership encourages a culture of knowledge dissemination and sharing, first of all by the leader himself because this is the agent that initiates such process by imposing his/her beliefs, values and assumptions to the work team (Prusak and Matson 2006; Senge 1990). This bases on the core idea that shared visions emerge from personal visions (Senge 1990), as so, building a learning organization requires a leader inspiring the vision of the learning organization itself: indeed, to be a learning organization has no value in itself, it must serve the broader aims of the organization (Davenport and Prusak 1998).

Subsequently, we can name communication, understood as the interaction between individuals within the organization that facilitates the free sharing of knowledge throughout time and at all levels, as another core characteristic of a learning organization. On the same logics, education shall as well be cited as another crucial aspect of a learning organization, based on the fact that it refers to the commitment by all organizational structures towards the leverage of continuous education opportunities, which is assumed as a foundation of the learning organization concept. In fact, and alluding to Cullen (1999) and Otala (1995), the learning organization develops and succeeds through a culture where individual development is considered as a priority, so that individuals are empowered to profit from jointly defined goals. Derived from the characteristics already mentioned, we can as well evidence culture as another chief aspect of a learning organization, based on the premise that such organizations shall gather efforts towards the development of a culture that supports, endorses and rewards learning as a vital part of the organization's enhancement. In fact, successful organizations habitually produce new knowledge, diffuse it all along the organization and integrate it in new processes, practices, products and services, reflecting a culture in which daily activities consubstantiate a learning opportunity for unceasing organizational improvement. This aspect is indeed transversal to the other above mentioned, in the first instance because culture begins with leadership, even though it does not extinguishes in it. Indeed, culture is the result of a group's accumulated learning, subsequently, the culture itself will later define the wanted leadership style.

All this can only be efficiently achieved through processes and infrastructures, being it management or technical ones, that sustain the learning organization. Indeed, organizational learning requires inquisitiveness and openness, thus, a willingness to challenge assumptions and tackle conventional wisdom. So, a successful learning organization sees a learning opportunity in every experiment and always thrives to revamp itself in order to malleably adjust, to embrace changes in its environment (Braham 1996; Senge 1990): hence, nuclear to a learning organization is the process of searching, assessing, evaluating and synthesizing evidence related to organizational issues, as learning organizations aim to catalyse the process of obtaining, exploiting and applying newly obtained knowledge (Quinn 1992).

Beyond these definitions, it is helpful to identify and operationalize the key functions and core tasks of a learning organization, keeping in mind that each function includes a complex of tasks: information gathering and problem solving, experimentation, learning from past, learning from promising practices, and transferring knowledge (Garvin 2000). This contribution benefited from Senge (1990), one of the first organizational researchers to expose the concepts that underlie the development of a learning organization: systems thinking (focusing on the organizational multiple relationships), personal mastery (process aimed at clarifying what is important), mental models (process aimed at clarifying and adjusting underlying assumptions), shared vision (agreeing on defined goals and course of action), and team learning (thinking insightfully, joining the individual and collective input, generating new learning). According to Senge (1990), the core of a learning organization work roots on the above five "learning disciplines", as they all consubstantiate into an interactive system, providing tools and methods that are applicable in the process of organizational knowledge creation, appropriation, application and transfer, so that each vector affords a valuable input to the enduring process of edifying an organisation that can indeed learn and capitalize on such knowledge (Gorelick 2005). Each of those vectors impacts the organizational dynamics on three levels—practices, principles, essences—developing and articulating in harmony to the three levels systematized by Van Gigch (1991).

Henceforth, organizational learning focuses on the process through which learning is developed in an organization (Yeo 2005): this assumption states that, although researchers and practitioners tend to use the key terms organizational learning and learning organization interchangeably, conceptual distinctions can be observed. Such dissimilarity was clearly addressed by Tsang (1997), intensely pointing out that organizational learning denotes the study of the organizational learning process, focusing on an academical approach. On the other hand, the learning organization is apprehended as an entity disposing of the capacity to effectively and efficiently learn, continuously adapting in face of the environment, sustaining and prospering in the long-term. As so, works in regards to learning organizations commonly intend at understanding the paths and processes to create and improve the organizational learning capacity, thus impelling a more practical impact and a performance agenda. Organizational learning entails the dynamic process of improving actions through better knowledge and understanding (Fiol and Lyles 1985), process which is based on knowledge conceived as a key strategic resource to organizational learning (Jerez-Gómez et al. 2005). Such process then implies a transversal approach through different organizational levels of action, all along the organizational structure (Crossan et al. 1999; Huber 1991). By its end, learning organizations configure an outcome: through other words, a learning organization is the normative facet of organizational learning, as it consists of the ultimate goal that an organization endeavours to accomplish, whereas organizational learning refers to the means through which a learning organization is achieved (Firestone and McElroy 2004). According to Gorelick (2005) and Kezar (2005), a learning organization tends to focus on external threats as the root for fostering learning, by its end, organizational learning consists of a continuous learning cycle which tends to shift the attention towards internal concerns for performance and learning, presenting them as a key component of human beings as organizational essential actors.

4 Knowledge Management

Hamel and Prahalad (1996) accentuated the relevance of people, knowledge, competencies and skills as organizational core values, advocating that knowledge is indeed the central component in the contemporary informational era. Indeed, knowledge assumes as an important resource for an organization, a basic source of competitive advantage as worker knowledge and intellectual capital have growingly been recognized as critical in organizational success (Gold et al. 2001; Conner and Prahalad 1996; Grant 1996; Jaworski and Kohli 1993), based on the premise that strategic learning aims to support future strategic initiatives that will, in turn, foster knowledge asymmetries possibly conducting to variations in organizational performance (Thomas et al. 2001). Thus, in the current paradigm the value in organizational contexts is attributed to intangible assets, rather than to traditional raw materials as in previous approaches (Tissen et al. 1998).

The notion of knowledge management is fairly recent, still developing once it arrived only around 1990s (Alavi and Denford 2011), foremost due to the input from relevant consultancy organizations seeking to capitalize on the massive potential of information technology, subsequent to a period of dissatisfaction with the prescriptions of re-engineering (Hammer and Champy 1993). Thus, impelling greatly from practitioners' concerns, knowledge management then as well achieved academic legitimacy as a driving force in the corporate world on the back of Nonaka's (1994) work. The idea derived from the neo-economic perspective of the strategic value of organizational knowledge, furthering it introducing informatically communication technologies and software in order to enable, in the quickest and simplest process possible, the generation, achievement, appropriation, sharing, storage and utilization of knowledge. Therefore, the conceptual logics grounding this construct follows the technical perspective of organizational learning as expanded by Huber (1991).

Among the definitions of knowledge management, Malhotra's (1998) is particularly important as the author postulates it configures the most critical question of organizational adaptation and survival due to present and growing changes in business environment. In summary, knowledge management consists of the process per which an organization generates values from its intellectual assets based on knowledge. Hence, knowledge management embraces the adoption of collective knowledge in order to achieve the organizational goals, consubstantiating a systematic effort in order to enable information and knowledge to grow, flow and create value (O'Dell and Hubert 2011), assuming an active and decisive role ensuring that people dispose of, appropriate and use the right knowledge at the right place and at the right time.

From here we can derive the assumption that the most important pillar in knowledge management is that knowledge is effectively and efficiently used on decision-making, because in its essence knowledge management concept presents the ability to attain the necessary information in short term, which will provide that all agents are able to bring the best decision about market scenarios, competitors, distribution channels, products, services and/or any other actions which are important for organizational sustainability (Gold et al. 2001; Shockley 2000). We then derive that with efficient knowledge management process, organizations will dispose of an effective input towards innovation, which may then foster organizational performance (Darroch 2005).

Via a wide-ranging review of knowledge management conceptual suggestions, Awad and Ghaziri (2004) summarized six vectors regularly used to define knowledge management, including the utilization of accessible knowledge both from outside and/or inside sources, diffusing knowledge throughout the organization, encouraging knowledge growth as an integral chief component of the organizational culture, codifying and exhibiting knowledge both in databases and/or documents, thus embedding and storing it, and assessing the value of knowledge assets and impact. As a synthesis, all these vectors build upon a foundation of information management, so that, at its essence, knowledge management entails organizational processes that strive towards the interaction of data and information, which then result in the improvement of information technologies, as well as on the recreation and innovation of human capacities.

Indeed, researchers have identified core aspects in the knowledge management process (Ivers 1998; Skyrme 1998; Teece 1998; Spender 1996; Leonard-Barton 1995), which we can systematize into four broad dimensions of process capability, namely acquisition, conversion, application and protection (Cui et al. 2005; Gold et al. 2001). As a synthesis, we can support that the widest approach towards this concept heritages on the premise that the knowledge management consists of a unity among three crucial components—people, processes and technology—developing through stages—knowledge creation, capture, storing, sharing and application—this way entailing its own life cycle.

Wiig et al. (1997) suggest that knowledge management entails, as core purposes, to enable the most efficiently possible decision making process and to recognise the maximum value of an organization's knowledge assets, in order to secure its sustained success. Hence, the chief concern and objective of knowledge management for an organization should be the focus on the long term, thus, to produce a learning organization capable of generating, determining, accessing, storing, transferring, appropriating and reconfiguring knowledge, capitalizing on the skills and expertise of its employees in order to create an organization that constitutes an entity which is more than the mere sum of its constituents. Furthering this argument, we can state that the first assumption basing the fact that organizations are interested in managing knowledge roots on the premise that organizational core competencies are built an developed on the skills, knowledge and experience of the people who do the work, the organizational agents, hence, they may not exist as systematized, organized and stored in a physical form (Manville and Foote 1996). This way, it is vital that organizations elaborate a strategy and design a subsequent process aimed at efficiently tap into this knowledge base in order to preserve and expand their core competencies, being aware that, and as stated previously, knowledge is triumphing as the driving force in the contemporary global economy, so that it becomes critical for an organization to access existing knowledge, as well as to create, appropriate, diffuse and capitalize on new knowledge. On the same line, Gold et al. (2001) examine the issue of effective knowledge management from the perspective of organizational capabilities, presenting a basis for understanding the competitive predisposition of an organization as it enters a program of knowledge management, suggesting that a knowledge infrastructure is required as an organizational precondition for effective knowledge management, and that infrastructure shall entail technology, structure and culture with a knowledge process architecture of acquisition, conversion, application and protection.

So, we can defend that knowledge management is experiential, driven by action-learning and primarily adopted by knowledge workers, being as well facilitated by organisational vision and support, involving continuous cycles of creativity and innovation. Acknowledging the assumption that controlling knowledge processes is problematic, based on the idea that knowledge is multi-faceted and complex, being both physical and mental, implicit and explicit, verbal and encoded, situated and abstract, distributed and individual, developing and static (Blackler 1995), this conceptual construct must then be contingent on the appropriate contexts in order to be efficient (Von Krogh et al. 2012; Wang and Ahmed 2003; Leonard-Barton and Swap 1999; Peres-Bustamante 1999; Nonaka and Takeuchi 1995). Subsequently, it must encompass management tools, practices and processes that will foster the processes of knowledge creation, transfer, sharing, dissemination and application in the organisation, once it is generally focussed on apprehending an organization's know-how and know-what via its conception, collection, storage, appropriation, spreading and application. Accordingly, knowledge management as a process and an approach refers to identifying, systematizing and connecting the organizational collective knowledge obtained through its (and its members) experience and competencies. Rooted on this assumption, the knowledge management conceptual construct has been sharply criticized by the social school of organizational learning, for example on the work of Brown and Duguid (2000): this perspective pointed out its unawareness of the social architecture deeply imprinted on the process of knowledge exchange within organizations. Aimed at filling in such gap, the social viewpoint has inserted and reconfigured technologies into the organizational context, based on the premise that it would enable sharing of tacit knowledge between its agents, positively impacting on flexible communication (McAfee 2006). Following such logics, Wiig et al. (1997) postulated that "management" infers that a defined asset has to be managed, and by extension, that that asset may be conceived as an object, which, by its definition, is habitually presumed to be tangible and concrete. However, knowledge is not configured as so, even though, it is passible to be measurable. By expansion if this argument, we state that organizational knowledge is also intangible as it defines the organization and is a reflection of the organizational culture. Even though, Von Krogh et al. (2012) coined the expression "knowledge enablement" considering that the term "management" when applied to knowledge dynamics integrated in the organizational context would be inadequate as it requires the capacity of controlling processes that, by its essence, may be uncontrollable and difficult to direct. However, Fleury (1997) argues that the term "management" would not be entirely inappropriate to describe and analyse organizational knowledge dynamics, once it involves other tools, processes, approaches and meanings that are beyond control, specifically planning, organising and evaluating.

Considering the premise that organizational knowledge is a strategic asset, from here we derive that the method used to implement a knowledge management system is critical. On such basis, Wiig (1997) acknowledged five strategies used by organizations in order to implement knowledge management systems, which differ taking into consideration the individual business and the organizations' unique needs, hence explaining that indeed some organizations pursue knowledge as a business strategy (the focus is on knowledge creation, apprehension, organization, sharing, use and transformation, at each point of action); others imprint a focus on intellectual asset management (namely patents, technologies, structural knowledge assets, customer relations, operations and management practices); others, by their end, concentrate on a personal knowledge asset accountability strategy (where each employee is responsible for his/her own knowledge related investments, within his/her area of accountability); another branch focusses on knowledge creation (privileging organizational learning, research, development, innovation and employee motivation); lastly, another set emphasises knowledge transfer (enhancing systemic approaches to diffuse and share knowledge).

As a synthesis of the above mentioned, we defend that the process of knowledge management can be understood from three levels: individual, group and organizational. The individual level comprises the dynamics of tacit and explicit knowledge, given the fact that, as individuals create information and obtain knowledge, it is then shared within the organization through social interactions creating new knowledge. Even though, knowledge sharing depends not only on the diffusion dynamics observed on the individual and team level, but also on the plethora of organizational factors underlying the successful implementation of a knowledge management system (Austin 2008). So, literature exhibits three major theoretical approaches to the conceptual construct and managerial applications of knowledge management (Poynder 1998): one advocates that knowledge management is predominantly an information technology topic; a second proposes that it concentrates on a human resources issue highlighting organizational culture and teamwork dynamics; a third encourages the development of processes to attain and access the organization's knowhow, which are, nevertheless, not required to involve the use of information technology.

As so, several benefits from knowledge management can be anticipated (Lank 1997): such process may boost employees to advance their performance and employability by expanding resources and providing them as immediately available, thus empowering such workforce to make and to process more intelligent decisions. As well, an effective knowledge management process may engender decreased levels of stress for employees trying to perform more tasks with fewer resources, as this workforce may then be able to spend less time seeking for information and expertise, which, by its turn, may enable such professionals to concentrate on their areas of expertise. This climaxes in the idea that a knowledge management process may then as well support organizations to convert into more competitive units by applying new knowledge to reduce costs, increase speed and satisfy customer needs (O'Dell et al. 1998). Such benefits to be accomplished, it is crucial to acknowledge that knowledge management must have a vision, strictly linked to organizational strategy and structure (Remus and Schub 2003; Mentzas 2001; Davenport et al. 1998; Nonaka and Takeuchi 1995): management has a central role on it (Jarrar 2002; Gore and Gore 1999), because knowledge management is mainly carried out by people, that way involving a combination of technical and human elements (Wong and Aspinwall 2004).

5 Clarifying Both Concepts: Matches and Mismatches

With a particular vigour as of the new millennium, massive efforts have been directed towards the study of the conceptual constructs and managerial approaches of learning organization, organizational learning and knowledge management. Indeed, if, as a title of example, Marquardt (1996) merely identified learning as the simplest means of knowledge acquisition, Nonaka (1991) exhibited the relevance of knowledge as the only ultimate source of organizational sustained competitive advantage. In a communion with such argumentation, Garvin (1993) postulated that a learning organization then consists of an organizational construct skilled on the process of generating, attaining, transferring and transforming knowledge, thus, at modifying behaviour to constantly reflect and incorporate new knowledge, impelling the organizational success towards the long-run. Hence, we can see that in the new millennium research boosted not only focusing on these concepts separately, highlighting their idiosyncrasies, but foremost examining their profound relationship, in fact systematizing that a learning organization complements knowledge management as it involves intricately incorporating what has been learned by individual organizational agents into the roots of a successful organization (Robbins and Judge 2009). Reverting the focus, but in a communion with such idea, Firestone and McElroy (2004) emphasize that knowledge management entails an organizational structure correspondent to the notion of the idealized learning organization outlook.

Hence, the concerns between the conceptual constructs are identical: according to Loermans (2002), one cannot exist without the other. Holsapple and Jones (2004), inspired by Senge (1990), believe that, deriving from the Globalization impact, the global deep interconnectedness and the growingly complex work dynamics, contemporary organizations, in order to remain competitive, are required to dispose of several strategic thinkers, which means, learning shall be a transversal process to all organizational agents, throughout the learning continuum from knowledge acquisition to its sharing, this way optimizing the decision-making process. Consequently, a learning organization consists of an organization where employees develop their competences and skills in order to accomplish desirable results, thus thriving new patterns of behaviour, also learning how to unceasingly generate and obtain new knowledge, impelling the organization into an incessant path of improvement. In those organizations we witness the implementation of knowledge management approaches as a means to achieve such organizational goals. Nevertheless, another trend in the literature postulates that the terminology evidently varies and understanding the relation between learning organization, organizational learning and knowledge management

5.3 Scope: System Versus Subsystem

Literature about knowledge management greatly discusses its influence on organizational learning: while some authors find these are cause and effect simultaneously, some others take organizational learning as a cause and knowledge management as an effect, or even opposite. So, authors expose distinct perspectives on the causal direction of this relationship and relatable impacts, presenting it either as a necessary pre-condition, or as a consequent reaction (Sue et al. 2003, 2004; Pemberton and Stonehouse 2000). Organizational learning tries to manage the organization's assets and to incessantly reload them with newly acquired knowledge. Thus, a learning organization refers to the whole organization, but as a component of the world, the organization must cooperate, through a network, with other sub-systems, a vital precondition for its survival. By its end, the ultimate goal of knowledge management is to formalize, store, distribute, share, coordinate and apply available knowledge throughout all the organization, leveraging the development and utilization of core skills and competences which stimulate outstanding performance.

In the view of Van Gigch's (1991) three levels of inquiry, a learning organization can be compared with the Reflecting level—"Why", by its end, knowledge management can be compared with the Diagnostic level—"What to do", based on the premise that a learning organization aims to support learning, which, as per its essence, requires changes in the existing knowledge. So, a learning organization involves an effective management of knowledge, and, by its turn, effective knowledge management consists on what to do to fulfil such requirement, because at its core such concept aims at supporting the distribution of knowledge, which requires, by logics, aiming to support learning. Nonetheless, a learning organization articulates on a much more complex level and through a broader approach than knowledge management, as it comprises work processes which must enable and empower learning and integrate every aspect of knowledge, so, it involves vectors beyond knowledge management. This indicates that work processes dynamized in a learning organization can be apprehended as a vital connection between the learning organization configuration and knowledge management approaches. This ground on the assumption that knowledge management process as a cycle, accordingly to Wielinga et al. (1997), Wiig (1997) and van der Spek and de Hoog (1995), articulated into four phases, namely to conceptualize, to reflect, to act and to review in regards to knowledge. Thus, in a synthetized assumption we postulate that a learning organization requires knowledge management, as knowledge management on the first instance assumes the existence and efficient functioning of a learning organization. In other words, a learning organization can be conceived as the system, which includes the subsystem encompassed by knowledge management. Indeed, knowledge management only understands organizational learning as a process which transforms knowledge, nevertheless, a learning organization excels at organizational learning and conceives it as far more referring to a collective cognitive process, even though its literature not explicitly discusses organizational memory (Aggestam 2006).

This argumentation develops in accordance to Senge's (1990) fundamental idea that system's thinking must be the conceptual cornerstone when discussing the construct of a learning organization, from here resulting the conceptual matches and mismatches to relatable notions. Consequently, we from here derive that a variation in knowledge management will trigger the potential to affect the organization as a whole unit, but on the other hand, an alteration in the organization, as for example on a core construct namely its culture, will by itself also affect knowledge management. This leads to the assumption that enabling organizations to effectively assume themselves as learning organizations requires introducing both organizational learning dynamics, as well as knowledge management approaches: so, we can postulate that the process required so that an organization effectively becomes a learning organization must take place on both levels. This is based on the fact that, without knowledge management, an organization will not be able to efficiently develop personal or group learning abilities (Su et al. 2004; Garratt 1990), once a learning organization indeed disposes of a culture which understands learning and knowledge sharing as a necessary and positive pre-requisite for organizational sustainability, but it is indeed knowledge management which can support this goal, because it assumes as a concrete means to operationalise the connection between the individual and organizational level. On an individual level, every organizational agent (employee) wants to contribute with their knowledge, skills, competences and experience, and also take part on other agents' knowledge through a constant sharing dynamics. Nonetheless, in order to capitalize on such knowledge, it must be efficiently and effectively integrated into the organization so that it can be assimilated, utilized and transformed in the most optimized way. Therefore, in order to develop learning abilities, the organization should surely design and implement optimal knowledge management process.

Another important aspect to clarify in regards to the conceptual distinction refers to the factors the theoretical constructs privilege in their analyses: if internal or external. The argumentation above supports the idea that knowledge management is performed in the organization, it involves a specific sub-system within the organization because it is evidently more attentive to internal factors inside the organization. So, when knowledge management literature cited external factors, it refers to aspects which the organization itself must manage: those concerns are often addressed in an implicit discourse. Nonetheless, discussions about how efficient knowledge management is are as well addressed, even though they are, on the other hand, explained as dependant on internal factors. So, we can defend that knowledge management discusses both external as internal factors in terms of organizational constraints, but the same does not happen when analysing the debates around the conceptual construct of learning organization. Indeed, when literature examining organizational learning discusses internal factors, the argumentation mostly flows in the view of the process towards meeting organizational internal demands. This is sustained on the premise that an organization is rooted on singular agents, thus, meeting internal demands in fact embraces meeting the concerns addressed by each employee, as each member possesses a self-concept of itself and of itself in relation to the organization. As a conclusion, we state that the organization's actions stem from these images and the interrelation they establish (Agyris and Schön 1996). In the sense that we detect system's thinking when discussing the dynamics of a learning organization, it can be argued that it as well implies this same approach to be applied, with the due specificities, to conceptual debates around knowledge management.

5.4 Agents: Individual Level Versus Collective Level

Learning, in an organizational context, occurs typically at a personal level, nonetheless, this initial and basic level remains insufficient to consider the unit as a learning organization, because, covered by such conceptual construct, learning cannot not be circumscribed to the personal level only, it should be further extended to reach the organizational one. Following this line, Scott-Kennel and von Batenburg (2012) reported that the organizations' ability to learn through employee's experience, fully acknowledging its depth and diversity, as well as the possible setbacks from there derived, is moderated by the efficiency of the mechanisms that enable application of knowledge to further internationalisation, being it influenced by internal tacit knowledge. Indeed, we can postulate that, to accomplish such goals, the personal struggle to acquire knowledge must be hampered by a facilitative corporate structure, so that we concretely assist to an effective promotion of learning at an organizational level by building knowledge propagation and transmission mechanisms throughout the whole structure (Loermans 2002; Christie and Sandelands 2000; Davenport and Prusak 1998).

Following the above thinking, when analysing the process agents on both conceptual constructs of organizational learning and knowledge management, we can see that individual knowledge consists of the pilot vector for organisational knowledge creation and development (Nonaka and Takeuchi 1995), and since information entails the raw material from which individual knowledge originates (Sarvary 1999), it entails the roots of the knowledge organisation and of organizational knowledge and learning (Choo 1998). It is noteworthy to highlight that individuals' knowledge ascends from the mixture of information, interpretation, reflection and experience within a given context (Davenport et al. 1998), consequently, in the learning organization we advocate, as a precondition, that individual knowledge is required to have form so that the possibility to use such knowledge can be permitted to the organizational agents. So, learning in organizations requires individual personal knowledge to be developed into a broader and more complex level, so that it is able to consubstantiate into information which other organizational members can spot, comprehend, apprehend, share and use (Jensen 2005). To create organisational knowledge, it is then clear that individual knowledge must be externalised (Nonaka and Takeuchi 1995), thus promoting group learning (Senge 1990) and knowledge propagation to all organisational levels. Hence, the transformation of individual knowledge into organisational knowledge occurs by a cadence of phases, namely via externalisation (conversion of tacit into explicit knowledge), internalisation (conversion of explicit into tacit knowledge), combination (conversion of explicit into explicit knowledge) and socialisation (conversion of tacit into tacit knowledge), keeping in mind that such process may take place from person to person, from a person to groups, or even via intra-groups dynamics (Nonaka and Takeuchi 1995).

In fact, a learning organization disposes of a culture understanding learning, knowledge creation and sharing as necessary and positive: on an individual level, every organizational agent thrives to contribute with his/her knowledge, skills, experience and capabilities to take part on other agents' knowledge, skills, experience and capabilities. To capitalize on such dynamics, in the organization's daily work this knowledge must be integrated: knowledge management indeed aims to support this, consisting of a concrete connection between the individual and the organizational level. This grounds on the premise that, once created, organisational knowledge depositories will contribute to leverage further individual knowledge, via a virtuous circle whose outcome must be the conversion of generated knowledge into organizational efficient actions. Nonetheless, the organisation must align strategies, policies, procedures and practices with the predefined organisational goals, acknowledging that individual knowledge, experience and competencies are difficult to control.

This way, we can defend that a shared view articulates as a mental map directing individuals by three allied areas: the world in which they live, the world in which they must live, and knowledge that needs to be settled to follow the trail between these two. So, from here we postulate that knowledge management is concentrated on implementing the opportunities so that all organizational agents are able to access, share, use and transform the organizational knowledge: it refers to the process resulting into an objective response to the concern that individual organizational agents must be capable to translate their learning into usable knowledge (Kezar 2005). Indeed, the purpose of knowledge management resides on the implementation of practices and procedures aimed at supplying the bases for organisational knowledge. Therefore, the knowledge management model entails a set of phases, according to Le Boterf (2000), through which we define the knowledge course, namely: meaning creation and shared vision of the knowledge development purposes; provision of information; induction to internal processing for individual knowledge creation; conversion of individual knowledge into group learning; knowledge dissemination to other organisational levels; and practical application of knowledge. This aspires at the collective use of organisational knowledge, encouraging the application of knowledge in problem-solving, specifically through product/service development and innovation, which is achieved as it as well fosters support during exposure to risks and leniency to errors. Therefore, the conceptual construct of learning organization involves the knowledge effective managing, and knowledge management, by its turn, requires a learning organization structure. So, knowledge management and learning organization concepts are different, however dependent on each other.

5.5 *Products: Individual Memory Versus Organizational Memory*

Concentrating on the premise that knowledge management aims to create value for the organisation as a whole entity (Wong and Aspinwall 2004), we advance that knowledge management enables individual learning which then contributes to the organizational learning processes. The core motivation and the vital goal to achieve through such dynamics is to leverage business values, impelling the organization towards the long term viability. Hence, when an individual learns something new, this organizational agent is creating and absorbing new knowledge, which will be stored in his/her personal memory, to be used as it is in its current configuration but also as a basis for the upcoming transformation towards new knowledge. So, we can argue that knowledge management processes will do the bridge to integrate, assimilate and operationalize individual learning into organizational knowledge. This fundaments on the premise that individual learning converts into organizational knowledge only when organizations work as effective and efficient holding environments for knowledge, hence, when organizations directly embody and signify knowledge in the sense that they personify strategies, tools and procedures aimed at performing complex tasks that might have been completed via alternative conducts (Agyris and Schön 1996).

Furthering this argumentation, we can defend that, in order to stimulate learning of other members in the organization, the knowledge must be stored in the organizational memory, which remains objectivised in books, documents, databases, etc. The organizational memory essentially corresponds to the knowledge management dynamics which result on the accumulation of knowledge outside people (as to say, organizational individual agents), thus, it denotes the stored organizational knowledge, intending to allow organizational knowledge sharing and reuse. Hence, we postulate that knowledge management must be adapted to business processes, integrated into every organizational critical business process (Remus and Schub 2003; Loermans 2002). Informational technologies are a prerequisite for effective knowledge management, the digital component is indeed currently a trend (Wong and Aspinwall 2004; Loermans 2002), in the perspective that organizational memory and the organizational technical domain are forcefully tied in regards to knowledge storage, dissemination and sharing: both notions regard knowledge as a product. Subsequently, from here we derive that knowledge, in the knowledge management conceptual lens, is an input to and/or a result from organizational learning.

In fact, a section of organizational knowledge is stored in digital-supported repositories, which configure as organizational memories, but the core dynamics takes place when that stored knowledge is diffused, transmitted and shared, therefore, apprehended, applicated and used (Aggestam 2006). That process indeed results into leveraging learning and possible new knowledge: knowledge management is precisely concerned with new knowledge as it supports organizational learning by both taking care of the result (which consists of the new knowledge), and transforming such result to be reachable for all individual agents in the organization. Therefore, we reach the assumption that, even though connected, both concepts distinguish because knowledge management can be conceived as a prerequisite for the effective and efficient articulation of a learning organization, based on the premise that a learning culture, vital and initial condition in a learning organization, only operationalizes if accurate and pertinent information is capable of flowing freely in a linked network. So, the logics reverts: knowledge management's efficiency is, in turn, dependent on the learning organization configuration.

6 Leveraging Organizational Performance Through Learning and Knowledge

Performance is a persistent topic in management literature, nonetheless, the concept of performance holds a broaden complex of interpretations (Subramaniam and Venkatraman 2001).

For the case of this work, it is noteworthy to highlight that the performance outcomes associated with organizational learning need to be carefully addressed, because most significantly, organizational performance, through organizational learning and knowledge management, benefits to manage organizational change (Yeo 2003). Indeed, we postulate that, due to the constant learning and, relatedly, to the superior knowledge management capabilities, organizations are then able to achieve and apply knowledge more effectively and efficiently both in their strategies, processes, practices and tools, all along the organizational structure and agents, which results in above-normal organizational performance levels.

In fact, when organizations develop increased knowledge management capabilities, they are capable of more effectively targeting and implementing marketing offers to meet customer needs (Jadad et al. 2000), so, feedback from customers, competitors and distribution/communication channels must be used to foster core competences, this way boosting the achievement of long-run supernormal profits. This bases on the assumption that an organization disposing of a strong organizational learning configures not only as a mere collector or depot of knowledge, but indeed assumes as a true processor of it, acting at a more refined level. Thus, we defend that learning and knowledge management affects organizational performance positively.

In fact, empirical studies around the learning organization conceptual construct support that system-level learning positively impels on organizational performance, as well, empirical research on organizational performance has empirically demonstrate significant correlations between the learning organization dimensions and the perceived financial and adaptive performance (Wetherington and Daniels 2013; Davis and Daley 2008; McHargue 2003; Ellinger et al. 2002). In more detail, the significant correlations between organizational performance and accounting measures around financial performance—specifically return on equity, market value added and net income per employee—provide sustenance to the connection between a learning organization and leveraging financial performance in both profit (Davis and Daley

2008; Ellinger et al. 2002) and non-profit vectors (Wetherington and Daniels 2013; McHargue 2003).

Yet, it still remains poorly understood which are the underlying mechanisms of how learning actually influences organizational performance, as well as how these behave and impact. Recent empirical studies using structural equation modelling report that learning organization dimensions positively affect aspects of organizational intangible performance, namely knowledge and adaptive performance (Kim et al. 2017; Kim and Kim 2016), which recommends that a learning culture enhances the design, implementation and, later, the capitalization on knowledge creation and innovation systems within the organization. Apropos, empirical research has as well been stating that system-level dimensions, such as system connections or strategic leadership for learning, steadily evidence higher correlations with organizational performance than individual or team dimensions, particularly knowledge performance (Watkins 2017). Moreover, this trend on recent empirical research around the linkage between superior organizational performance and organizational core vectors has disclosed that a learning organization may ultimately enhance tangible financial performance indirectly through intangible knowledge performance, connection articulated via a mediated relationship. Also interesting to reference is the emerging research that growingly concentrates at moving beyond traditional associations and contexts elucidative of organizational performance to now explore the nature of performance in highly relational, interorganizational, global and non-profit collaborations (Watkins and Kim 2018).

7 Conclusion

The present work roots on the premise that the concepts of learning organization, organizational learning and knowledge management are obviously related, although not identical. Indeed, they have been used to define each other, often overlapping despite conceptual idiosyncrasies, impelling empirical analysis to become ambiguous. Through literature review, we aimed at filling in this gap, providing further clarification on the specificities of each conceptual construct. We argue that clear differences can be presented when comparatively analysing such conceptual constructs, which we summarized into five core vectors operating at the level of the conceptual approach (interest of academics vs. interest of practitioners), the goal (process vs. result), the scope (system vs. subsystem, the process (cause vs. effect), the agents (individual level vs. collective level) and the products (individual memory vs. organizational memory).

On the first vector, we advocate that the conceptual constructs originate and develop at different levels of abstraction (Aggestam 2006): literature on organizational knowledge greatly adopts a philosophical angle, focusing on understanding and conceptualizing the essence of knowledge within organizations, so that often the approaches are incisive on the distinctions between individual and organizational knowledge, the impact and usefulness of the distinction between tacit and explicit knowledge, how and which knowledge is shared, how knowledge is stored, or even debating whether and under which conditions may knowledge configure as a strategic organizational advantage or asset. On the other hand, literature on organizational management generally adopts a technical approach aimed at creating effective tools, practices and processes of accessing, measuring, appropriating, disseminating, transforming, codifying, storing and leveraging knowledge in order to enhance organizational performance. Thus, literature on organizational learning has mainly derived from academical interest, while literature about knowledge management originated greatly from practitioners concerns and inputs.

In regards to the second vector, we can state that a learning organization focuses on the learning process, and knowledge management, by its end, concentrates on the output of the learning process, as knowledge management's aim is to create increased organizational value (Loermans 2002). Knowledge management then embodies a vital process in the learning organization, as it encompasses acquiring, elucidating and communicating mission-specific professional expertise to organizational participants through a timely, focused and pertinent method (King 1999). Consequently, knowledge management emphasises the key role of tacit knowledge embedded in organizational processes, rooting and leveraging professional expertise, while a learning organization aims at endorsing the acquisition and dissemination of a wider range of information and general knowledge that might influence future opportunities for the organization, impelling the organizational performance towards the long-run sustainability (Garvin 1993).

The third vector address the fact that literature about knowledge management greatly discusses its different influence on organizational learning: while some authors find these two focuses are cause and effect simultaneously, others take organizational learning as a cause and, subsequently, knowledge management as an effect, or even opposite (Su et al. 2003, 2004). Based of Senge's (1990) argumentation, and as stated on the point above, we adopt the viewpoint considering a learning organization as the system, which, by its end, unavoidably contains the subsystem formed by knowledge management, in the way that a modification in knowledge management dynamics will disturb the organization as a whole unified entity, on the other hand, a transformation in the organization itself will as well affect knowledge management approaches (Su et al. 2004; Garratt 1990), once a learning organization disposes of a culture which understands learning and knowledge sharing as a necessary and positive pre-requisite for organizational sustainability and success, but it is indeed knowledge management dynamics which can support this goal, because it assumes as a concrete means to operationalise the connection between the individual and the organizational level. So, this vector clearly postulates that organizational learning tries to manage the organization's assets and to incessantly reload them with newly acquired knowledge: thus, a learning organization refers to the whole organization, the system which includes the subsystem encompassed by knowledge management. Hence, the learning organization, as a conceptual construct, refers to a more complex scope than knowledge management, as it concentrates on encouraging the implementation of a learning culture and the development of structures supportive of learning processes.

The forth vector defends that learning, in an organizational context, occurs typically at a personal level, nonetheless, this initial and basic level remains insufficient to consider the unit as a learning organization, because, covered by such conceptual construct, learning cannot not be circumscribed to the personal level, it should indeed be further extended to reach the organizational level. So, the organizations' ability to learn through employee's experience, fully acknowledging its depth and diversity, as well as the possible setbacks from there derived, is moderated by the efficiency of the mechanisms that enable the application of knowledge to further internationalisation: the personal struggle to acquire knowledge must be hampered by a facilitative corporate structure, so that we assist to an effective promotion of learning by building knowledge propagation and transmission mechanisms throughout the whole organization (Loermans 2002; Christie and Sandelands 2000; Davenport and Prusak 1998). Hence, the transformation of individual knowledge into organisational knowledge occurs by a cadence of phases, namely via externalisation (conversion of tacit into explicit knowledge), internalisation (conversion of explicit into tacit knowledge), combination (conversion of explicit into explicit knowledge) and socialisation (conversion of tacit into tacit knowledge), keeping in mind that such process may take place from person to person, from a person to groups, or even via intra-groups dynamics (Nonaka and Takeuchi 1995). To capitalize on such dynamics, this knowledge must be integrated in the organization's daily work: knowledge management indeed aims at supporting this, consisting of a concrete connection between the individual and the organizational level.

The fifth vector argues that, concentrating on the premise that knowledge management aims to create value for the organisation as a whole entity (Wong and Aspinwall 2004), knowledge management enables individual learning which then contributes to the organizational learning processes. Hence, when an individual learns something new, this organizational agent is creating and absorbing new knowledge, which will be stored in his/her personal memory, to be used as it is in its current configuration but also as a basis for the upcoming transformation towards new knowledge. So, knowledge management processes will do the bridge to integrate, assimilate and operationalize individual learning into organizational knowledge, as an outcome passing on from the individual memory to the collective memory (Agyris and Schön 1996).

Furthering this argumentation, with the present work we also address the challenge for future research to develop an assessment methodology, accordingly tested and validated, to afford management with the ability to effectively access the organizational progress both in implementing knowledge management and in the pursuit of a learning organization, even though, and truth to be said, to accomplish this measurement objective it is a *sine qua non* condition a theoretical conceptual framework first to be consolidated, which will not only offer the basis for the development of measurement tools and procedures, but will also allow and enhance the development of future research hypotheses that can then be accordingly tested. As with the present work we aimed at providing useful insights towards this aim, filling in this gap in the literature, we hereby clearly state its relevance.

References

Ackoff, R. (1997). Systems, messes and interactive planning. *The Societal Engagement of Social Science, 3,* 417–438.

Adams, G., & Lamont, B. (2003). Knowledge management systems and developing sustainable competitive advantage. *Journal of Knowledge Management, 7*(2), 142–154.

Aggestam, L. (2006). Learning organization or knowledge management—which came first, the chicken or the egg. *Information Technology and Control, 35*(3), 295–302.

Aktharasha, U., & Anisa, H. (2011). Knowledge management system and learning organization: an empirical study in an engineering organization. *The IUP Journal of Knowledge Management, 9*(2), 26–43.

Alavi, M., & Denford, J. (2011). *Knowledge management: Process, practice and web 2.0. Handbook of organizational learning and knowledge management.* West Sussex: Wiley.

Allee, V. (1997). 12 principles of knowledge management. *Training & Development, 51*(11), 71–74.

Argyris, C., & Schon, D. (1996). *Organizational Learning II: Theory, method and practice.* Massachusetts: Addison-Wesley.

Attewell, P. (1992). Technology diffusion and organizational learning: The case of business computing. *Organization Science, 3*(1), 1–19.

Austin, M. (2008). Strategies for transforming human service organizations into learning organizations: Knowledge management and the transfer of learning. *Journal of Evidence-Based Social Work, 5*(3/4), 569–596.

Awad, E., & Ghaziri, H. (2004). *Knowledge management*. New Jersey: Pearson Education Ltd.

Bass, B., & Avolio, B. (1993). Transformational leadership and organizational culture. *Public Administration Quarterly, 17*(Spring), 112–121.

Bierly, P., Kessler, E., & Christensen, E. (2000). Organizational learning, knowledge and wisdom. *Journal of Organizational Change Management, 13*(6), 595–618.

Blackler, F. (1995). Knowledge, knowledge work and organizations: An overview and interpretation. *Organization Studies, 16*(6), 1021–1046.

Blodgood, J., & Salisbury, W. (2001). Understanding the influence of organizational change strategies on information technology and knowledge management strategies. *Decision Support Systems, 31*(1), 55–69.

Bollinger, A., & Smith, R. (2001). Managing organizational knowledge as a strategic asset. *Journal of Knowledge Management, 5*(1), 8–18.

Braham, J. (1996). *Creating a learning organization.* Simi Valley: Kogan Page.

Brown, J., & Duguid, P. (2000). *The Social life of information.* Boston: Harvard Business School Press.

Cangelosi, V., & William, D. (1965). Organizational learning: Observations toward a theory. *Administrative Science Quarterly, 10*(2), 175–203.

Chinowsky, P., & Carrillo, P. (2007). Knowledge management to learning organization connection. *Journal of Management in Engineering, 23*(3), 122–130.

Choo, C. (1998). *The knowing organization.* New York: Oxford University Press.

Choo, A., Linderman, K., & Schroeder, R. (2007). Method and context perspectives on learning and knowledge creation in quality management. *Journal of Operations Management, 25*(4), 918–931.

Christie, A., & Sandelands, E. (2000). The knowledge harvest: Ensuring you reap what you sow. *Journal of Workplace Learning, 12*(3), 83–89.

Chuang, S. (2004). A resource-based perspective on knowledge management capability and competitive advantage: An empirical investigation. *Expert Systems with Applications, 27*(3), 459–465.

Cole-Gomolski, B. (1999). Knowledge 'czars' fall from grace. *Computerworld, 33*(1), 1–13.

Conner, K., & Prahalad, C. (1996). A resource-based theory of the firm: Knowledge versus opportunism. *Organization Science, 7*(5), 477–501.

Coopey, J. (1995). The learning organization, power, politics and ideology introduction. *Management Learning, 26*(2), 193–213.

Coopey, J., & Burgoyne, J. (2000). Politics and organizational learning. *Journal of Management Studies, 37*(6), 869–886.

Crossan, M., Lane, H., & White, R. (1999). An organizational learning framework: From intuition to institution. *Academy of Management Review, 24*(3), 522–537.

Cui, A., Griffith, A., & Cavusgil, S. (2005). The influence of competitive intensity and market dynamism on knowledge management capabilities of MNC subsidiaries. *Journal of International Marketing, 13*(3), 32–53.

Cullen, J. (1999). Socially constructed learning: A commentary on the concept of the learning organisation. *The Learning Organization: An International Journal, 6*(1), 45–52.

Cunningham, J., & Gerrard, P. (2000). Characteristics of well-performing organizations in Singapore. *Singapore Management Review, 22*(1), 35–64.

Cyert, R., & March, J. (1963). *A behavioural theory of the firm*. Englewood Cliffs: Prentice Hall.

Daft, R., & Weick, K. (1984). Toward a model of organizations as interpretation systems. *Academy of Management Review, 9*(2), 284–295.

Darroch, J. (2005). Knowledge management, innovation and firm performance. *Journal of Knowledge Management, 9*(3), 101–115.

Davenport, T., & Prusak, L. (1998). *Working knowledge: How organizations manage what they know*. Massachusetts: Harvard Business Press.

Davenport, T., DeLong, D., & Beers, M. (1998). Successful knowledge work processes. *Sloan Management Review, 39*(2), 43–57.

Davis, D., & Daley, B. (2008). The learning organization and its dimensions as key factors in firms' performance. *Human Resource Development International, 11*(1), 51–66.

DiBella, A., Nevis, E., & Gould, J. (1996). Understanding organizational learning capability. *Journal of Management Studies, 33*(3), 361–379.

Drucker, P. (1992). *Managing for the future: The 1990s and beyond*. New York: Truman Talley Books/Plume.

Edmondson, A., & Moingeon, B. (1998). From organizational learning to the learning organization. *Management Learning, 29*(1), 5–20.

Ellinger, A., Ellinger, A., Yang, B., & Howton, S. (2002). The relationship between the learning organization concept and firms' financial performance: An empirical assessment. *Human Resource Development Quarterly, 13*(1), 5–22.

Epple, D., Argote, L., & Devadas, R. (1991). Organizational learning curves: A method for investigating intra-plant transfer of knowledge acquired through learning by doing. *Organization Science, 2*(1), 58–70.

Fiol, C., & Lyles, M. (1985). Organizational learning. *Academy of Management Review, 10*(4), 803–813.

Firestone, J., & McElroy, M. (2004). Organizational learning and knowledge management: The relationship. *The Learning Organization, 11*(2), 177–184.

Fleury, S. (1997). Science in social studies: Reclaiming science for social knowledge. In E. W. Ross (Ed.), *The social studies curriculum: Purposes, problems, and possibilities* (pp. 165–182). New York: SUNY Press.

Garratt, B. (1990). *Creating a learning organization: A guide to leadership, learning, and development*. New York: Simon & Schuster.

Garratt, B. (1987). Learning is the core of organisational survival: action learning is the key integrating process. *Journal of Management Development, 6*(2), 38–44.

Garvin, D. (2000). *Learning in action: A guide to putting the learning organization to work*. Massachusetts: Harvard Business Press.

Garvin, D. (1998). *Building a learning organization in Harvard Business Review on knowledge management*. Massachusetts: Harvard Business Press.

Garvin, D. (1993). Building a learning organization. *Harvard Business Review, 71*(4), 78–91.

Garvin, D.A., Edmondson, A.C., & Gino, F. (2008). Is Yours a Learning Organization?—Tool Kit. *Harvard Business Review,* 1–10. https://www.hbr.org.

Goh, S., & Richards, G. (1997). Benchmarking the learning capability of organizations. *European Management Journal, 15*(5), 575–583.
Gold, A., Malhotra, A., & Segars, A. (2001). Knowledge management: An organizational capabilities perspective. *Journal of Management Information Systems, 18*(1), 185–214.
Gopher, D., Well, M., & Bareket, T. (1994). Transfer of skill from a computer game trainer to flight. *Human Factors, 36*(3), 387–405.
Gore, C., & Gore, E. (1999). Knowledge management: The way forward. *Total Quality Management, 10*(4/5), 554–560.
Gorelick, C. (2005). Organizational learning vs the learning organization: a conversation with a practitioner. *The Learning Organization, 12*(4), 383–388.
Gourlay, S. (2001). Knowledge management and HRD. *Human Resource Development International, 4*(1), 27–46.
Grant, R. (1996). Toward a knowledge-based theory of the firm. *Strategic Management Journal, 17*(2), 109–122.
Hamel, G., & Prahalad, C. (1996). Competing in the new economy: Managing out of bounds. *Strategic Management Journal, 17*(3), 237–242.
Hammer, M., & Champy, J. (1993). *Reengineering the Corporation. A manifesto for business revolution*. London: Nicholas Brealey.
Harari, O. (1997). Flood your organization with knowledge. *Management Review, 86*(10), 33–37.
Harung, H. (1996). A world leading learning organization: A case study of Tomra Systems, Olso, Norway. *The Learning Organization, 3*(4), 22–34.
Hedberg, B. (1981). How organizations learn and unlearn. *Handbook of organizational design, 1*, 3–27.
Hendriks, P., & Vriens, D. J. (1999). Knowledge-based systems and knowledge management: friends or foes? *Information & Management, 35*(2), 113–125.
Holsapple, C., & Jones, K. (2004). Exploring primary activities of the knowledge chain. *Knowledge and Process Management, 11*(3), 155–174.
Huber, G. (1991). Organizational learning: The contributing processes and the literatures. *Organization Science, 2*(1), 88–115.
Ivers, J. (1998). Bringing out brilliance: enabling knowledge creation in the Notes/Domino environment. *Enterprise Solutions, 10*(November/December), 24–27.
Jadad, A., Haynes, R., Hunt, D., & Browman, G. (2000). The Internet and evidence-based decision-making: a needed synergy for efficient knowledge management in health care. *CMAJ, 162*(3), 362–365.
Jamali, D., & Sidani, Y. (2008). Learning organizations: diagnosis and measurement in a developing country context. *The Learning Organization, 15*(1), 58–74.
Jarrar, Y. (2002). Knowledge management: Learning for organisational experience. *Managerial Auditing Journal, 17*(6), 322–328.
Jaworski, B., & Kohli, A. (1996). Market orientation: review, refinement, and roadmap. *Journal of Market-Focused Management, 1*(2), 119–135.
Jensen, P. (2005). A contextual theory of learning and the learning organization. *Knowledge and Process Management, 12*(1), 53–64.
Jerez-Gomez, P., Cespedes-Lorente, J., & ValleCabrera, R. (2005). Organizational learning capability: a proposal of measurement. *Journal of Business Research, 58*(6), 712–725.
Kezar, A. (2005). What campuses need to know about organizational learning and the learning organization. *New Directions for Higher Education, 131*(Fall), 7–22.
Kim, S., & Kim, S. (2016). A multi-criteria approach toward discovering killer IoT application in Korea. *Technological Forecasting and Social Change, 102*, 143–155.
Kim, K., Watkins, K., & Lu, Z. (2017). The impact of a learning organization on performance. *European Journal of Training and Development, 41*(2), 177–193.
King, W. (1999). Integrating knowledge management into IS strategy. *Information Systems Management, 16*, 70–72.

Kolb, D. (1984). *Experiential learning: Experience as the source of learning and development*. Englewood Cliffs: Prentice Hall.

Lank, E. (1997). Leveraging invisible assets: the human factor. *Long Range Planning, 30*(3), 406–412.

Lawrence, T., Mauws, M., Dyck, B., & Kleysen, R. (2005). The politics of organizational learning: Integrating power into the 4I framework. *Academy of Management Review, 30*(1), 180–191.

Le Boterf, G. (2000). *Compétence et Navigation Professionnelle*. Paris: Éditions d'Organization.

Leonard-Barton, D. (1995). *Wellspring of knowledge*. Massachusetts: Harvard Business School Press.

Leonard-Barton, D. (1988). Implementation as mutual adaptation of technology and organization. *Research Policy, 17*, 251–67.

Leonard-Barton, D., & Swap, W. (1999). *When sparks fly*. Massachusetts: Harvard Business School Press.

Liao, S., & Wu, C. (2010). System perspective of knowledge management, organizational learning, and organizational innovation. *Expert Systems with Applications, 37*(2), 1096–1103.

Loermans, J. (2002). Synergizing the learning organization and knowledge management. *Journal of Knowledge Management, 6*(3), 285–294.

Lustri, D., Miura, I., & Takahashi, S. (2007). Knowledge management model: practical application for competency development. *The Learning Organization, 14*(2), 186–202.

MacLeod, M. (1999). Softworld'99: Knowledge management and how it can help streamline the supply chain. *Supply Management, 4*, 38–39.

Malhotra, Y. (1998). Tooks@ work: Deciphering the knowledge management hype. *The Journal for Quality and Participation, 21*(4), 58–60.

Manville, B., & Foote, N. (1996). Strategy as if knowledge mattered. *Fast Company, 2*(1), 66–67.

March, J. (1991). Exploration and exploitation in organizational learning. *Organization Science, 2*(1), 71–87.

Marquardt, M. (1996). *Building the learning organization*. New York: McGraw-Hill.

McAfee, A. (2006). Enterprise 2.0: the dawn of emergent collaboration. *MIT Sloan Management Review, 47*(3), 1–28.

McHargue, S. (2003). Learning for performance in nonprofit organizations. *Advances in Developing Human Resources, 5*(2), 196–204.

Mentzas, G. (2001). An holistic approach to realizing the full value of your knowledge assets. *Knowledge Management Review, 4*(3), 10–11.

Müller, J., & Stocker, A. (2011). Enterprise microblogging for advanced knowledge sharing: The references@ BT case study. *Journal of Universal Computer Science, 17*(4), 532–547.

Nonaka, I. (1994). A dynamic theory of organizational knowledge creation. *Organization Science, 5*(1), 14–37.

Nonaka, I. (1991). *The knowledge-creating company* (pp. 96–104). Nov-Dec: Harvard Bus Review.

Nonaka, I., & Takeuchi, H. (1995). *The knowledge-creating company*. New York: Oxford University Press.

Nonaka, L., Takeuchi, H., & Umemoto, K. (1996). A theory of organizational knowledge creation. *International Journal of Technology Management, 11*(7/8), 833–845.

O'Dell, C., & Hubert, C. (2011). *The new edge in knowledge: How knowledge management is changing the way we do business*. New Jersey: Wiley.

O'Dell, C., O'dell, C., Grayson, C., & Essaides, N. (1998). *If only we knew what we know: The transfer of internal knowledge and best practice*. New York: Simon and Schuster.

Otala, M. (1995). The learning organization: Theory into practice. *Industry and Higher Education, 9*(3), 157–164.

Pedler, M., Burgoyne, J., & Boydell, T. (1991). *The learning company*. Maidenhead: McGraw-Hill.

Pedler, M., Boydell, T., & Burgoyne, J. (1989). Towards the learning company. *Management Education and Development, 20*(1), 1–8.

Pemberton, J., & Stonehouse, G. (2000). Organisational learning and knowledge assets–an essential partnership. *The Learning Organization, 7*(4), 184–194.

Peres-Bustamante, G. (1999). Knowledge management in agile innovative organizations. *Journal of Knowledge Management, 3*(1), 6–17.

Poynder, R. (1998). Getting to the nuts and bolts of knowledge management. *Innovation World Review, 20,* 35–57.

Prusak, L., & Matson, E. (2006). *Knowledge management and organizational learning: A reader*. Oxford: Oxford University Press.

Quinn, J. (1992). *Intelligent enterprise: A knowledge and service based paradigm for industry.* New York: Free Press.

Remus, U., & Schub, S. (2003). A blueprint for the Implementation of Process-oriented Knowledge Managament. *Knowledge and Process Management, 10*(4), 237–253.

Robbins, S., & Judge, T. (2009). *Organizational Behavior* (12th ed.). Englewood Cliffs: Prentice Hall.

Sarvary, M. (1999). Knowledge management and competition in the consulting industry. *California Management Review, 41*(2), 95–107.

Schein, E. (2004). *Organizational culture and leadership* (3rd ed.). New Jersey: Wiley.

Schein, E. (1997). *A conceptual model for managed culture change. Organisational culture and leadership* (2nd ed.). San Francisco: Jossey-Bass.

Schein, E. (1993). On dialogue, culture and organizational learning. *Organizational Dymanics, 22*(2), 40–52.

Scott-Kennel, J., & von Batenburg, Z. (2012). The role of knowledge and learning in the internationalisation of professional service firms. *The Service Industries Journal, 32*(10), 1667–1690.

Senge, P. (1990). *The fifth discipline: the art and practice of the learning organization.* New York: Doubleday/Currency.

Sethi, V., & King, W. (1994). Development of measures to assess the extent to which an information technology application provides competitive advantage. *Management Science, 40*(12), 1601–1627.

Shockley, W. (2000). Planning for knowledge management. *Quality Progress, 33*(3), 57–64.

Sicilia, M.-A., & Lytras, M. (2005). The semantic learning organization. *The Learning Organization, 12*(5), 402–410.

Sicilia, M.-A., Lytras, M., Rodríguez, E., & García-Barriocanal, E. (2006). Integrating descriptions of knowledge management learning activities into large ontological structures: A case study. *Data & Knowledge Engineering, 57*(2), 111–121.

Simon, Y. (1991). *Practical knowledge.* New York: Fordham University Press.

Skyrme, D. (1998). Knowledge management solutions-the IT contribution. *ACM SIGGROUP Bulletin, 19*(1), 34–39.

Snell, R., & Chak, A. (1998). The learning organization: Learning and empowerment for whom? *Management Learning, 29*(3), 337–364.

Solomon, J. (1994). Knowledge, values and the public choice of science knowledge. *STS Education: International Perspectives on Reform*, 99–110.

Spender, J. (1996). Organizational knowledge, learning and memory: three concepts in search of a theory. *Journal of Organizational Change Management, 9*(1), 63–78.

Spender, J. (2006). Getting value from knowledge management. *The TQM Magazine, 18*(3), 238–254.

Spiegler, I. (2000). Knowledge management: a new idea or a recycled concept? *Communications of the Association for Information Systems, 3*(14). Retrieved September 03, 2020, from http://cais.aisnet.org/articles/3–14/default.asp? iew=html&x=27&y=9

Su, K., Huang, L., & Hsieh, H. (2004). The development of a knowledge flow paradigm in engineering education: empirical research in Taiwanese Universities. *World Transactions on Engineering and Technology Education, 3*(1), 125–128.

Su, K., Hsieh, H., & Liu, K. (2003). Application of the learning organization paradigm in engineering education: An empirical research of universities in Taiwan. *World Transactions on Engineering and Technology Education, 2*(2), 285–290.

Subramaniam, M., & Venkatraman, N. (2001). Determinants of transnational new product development capability: Testing the influence of transferring and deploying tacit overseas knowledge. *Strategic Management Journal, 22*(4), 359–378.

Sveiby, K. (1997). *The new organizational wealth: Managing & measuring knowledge-based assets.* Oakland: Berrett-Koehler Publishers.

Teece, D. (1998). Capturing value from knowledge assets: The new economy, markets for know-how, and intangible assets. *California Management Review, 40*(3), 55–79.

Thomas, J., Sussman, S., & Henderson, J. (2001). Understanding "strategic learning": Linking organizational learning, knowledge management, and sensemaking. *Organization Science, 12*(3), 331–345.

Thomas, K., & Allen, S. (2006). The learning organisation: A meta-analysis of themes in literature. *The Learning Organization, 13*(2), 123–139.

Thornburg, L. (1994). Accounting for knowledge. *HR Magazine, 39*(10), 50–56.

Tissen, R., Andriessen, D., & Lekanne, F. (1998). *Valuebased knowledge management: Creating the 21st century company: knowledge intensive, people rich.* Amsterdam: Addison Wesley Longman.

Tsang, E. (1997). Organizational learning and the learning organization: a dichotomy between descriptive and prescriptive research. *Human Relations, 50*(1), 73–89.

Tuomi, I. (2000). Data is more than knowledge: Implications of the reversed knowledge hierarchy for knowledge management and organizational memory. *Journal of Management Information Systems, 16*(3), 103–117.

Van der Spek, R., & De Hoog, R. (1995). A framework for a knowledge management methodology. *Knowledge management methods* (pp. 379–393). Arlington: Schema Press.

Van Gigch, J. (1991). *System design modeling and metamodeling.* New York: Plenum Press.

Von Krogh, G., Nonaka, I., & Rechsteiner, L. (2012). Leadership in organizational knowledge creation: A review and framework. *Journal of Management Studies, 49*(1), 240–277.

Wah, L. (1999). Making knowledge stick. *Management Review, May*, 24–29.

Wang, C., & Ahmed, P. (2003). Structure and structural dimensions for knowledge-based organizations. *Measuring Business Excellence, 7*(1), 51–62.

Watkins, K. (2017). Defining and creating organizational knowledge performance. *Educar, 53*(1), 211–226.

Watkins, K., & Kim, K. (2018). Current status and promising directions for research on the learning organization. *Human Resource Development Quarterly, 29*(1), 15–29.

Wetherington, J., & Daniels, M. (2013). The relationship between learning organization dimensions and performance in the nonprofit sector. *Journal for Nonprofit Management, 16*(1), 90–107.

Wielinga, B., Sandberg, J., & Schreiber, G. (1997). Methods and techniques for knowledge management: what has knowledge engineering to offer? *Expert Systems with Applications, 13*(1), 73–84.

Wiig, K. (1997). Knowledge management: an introduction and perspective. *Journal of knowledge Management, 1*(1), 6–14.

Wiig, K., de Hoog, R., & Van Der Spek, R. (1997). Supporting knowledge management: a selection of methods and techniques. *Expert Systems with Applications, 13*(1), 15–27.

Wong, K., & Aspinwall, E. (2004). Knowledge management implementation frameworks: A review. *Knowledge and Process Management, 11*(2), 93–104.

Yeo, R. (2003). *Defining science: William Whewell, natural knowledge and public debate in early Victorian Britain* (Vol. 27). Cambridge: Cambridge University Press.

Yeo, R. (2005). Revisiting the roots of learning organization. A synthesis of the learning organization literature. *The Learning Organization, 12*(4), 368–382.

Index

C. Machado and J. P. Davim (eds.), *Knowledge Management and Learning Organizations*,
https://doi.org/10.1007/978-3-030-71079-8